Environmental Impact Assessment

AF799454

Rachna Bhateria · Mona Sharma · Rimmy Singh · Sumit Kumar

Environmental Impact Assessment

A Journey to Sustainable Development

Springer

Rachna Bhateria
Department of Environmental Sciences
Maharshi Dayanand University
Rohtak, Haryana, India

Mona Sharma
Department of Environmental Studies
Central University of Haryana
Mahendragarh, Haryana, India

Rimmy Singh
Department of Applied Science
DPG Institute of Technology
and Management
Gurugram, Haryana, India

Sumit Kumar
Department of Environmental Science
and Technology
Guru Jambheshwar University of Science
and Technology
Hisar, Haryana, India

ISBN 978-3-031-66796-1 ISBN 978-3-031-66797-8 (eBook)
https://doi.org/10.1007/978-3-031-66797-8

© The Editor(s) (if applicable) and The Author(s), under exclusive license to Springer Nature Switzerland AG 2024

This work is subject to copyright. All rights are solely and exclusively licensed by the Publisher, whether the whole or part of the material is concerned, specifically the rights of translation, reprinting, reuse of illustrations, recitation, broadcasting, reproduction on microfilms or in any other physical way, and transmission or information storage and retrieval, electronic adaptation, computer software, or by similar or dissimilar methodology now known or hereafter developed.
The use of general descriptive names, registered names, trademarks, service marks, etc. in this publication does not imply, even in the absence of a specific statement, that such names are exempt from the relevant protective laws and regulations and therefore free for general use.
The publisher, the authors and the editors are safe to assume that the advice and information in this book are believed to be true and accurate at the date of publication. Neither the publisher nor the authors or the editors give a warranty, expressed or implied, with respect to the material contained herein or for any errors or omissions that may have been made. The publisher remains neutral with regard to jurisdictional claims in published maps and institutional affiliations.

This Springer imprint is published by the registered company Springer Nature Switzerland AG
The registered company address is: Gewerbestrasse 11, 6330 Cham, Switzerland

If disposing of this product, please recycle the paper.

Foreword

As Vice Chancellor of Central University of Haryana, it is my steadfast commitment to nurture a generation of environmentally aware and responsible individuals. I am pleased to introduce the textbook titled *Environmental Impact Assessment: A Journey to Sustainable Development*, authored by Rachna Bhateria, Mona Sharma, Rimmy Singh, and Sumit Kumar. This book endeavors to equip the next generation with essential expertise in Environmental Impact Assessment (EIA).

This book is a timely contribution to the environmental discourse, providing a clear and practical guide for navigating the EIA process. The authors' expertise is evident throughout, offering structured insights and a comprehensive understanding of EIA, empowering readers to actively participate in shaping a sustainable future.

The book simplifies the complexities of EIA into digestible chapters, guiding readers through each stage of the process with clarity and precision. It goes beyond theory, providing practical tools, methodologies, and real-world examples to conduct EIA studies effectively and contribute to environmental decision-making. By covering topics like Strategic Environmental Assessment (SEA) and Environmental Risk Assessment (ERA) alongside traditional EIA, the book offers a holistic perspective on environmental impact analysis.

The book explores nuances of EIA regulations and relevant notifications, making it invaluable for Indian professionals, students, and policymakers.

This book is particularly valuable for:

Students: Building a strong foundation in EIA principles and practices for successful careers in environmental fields. Professionals: Enhancing their EIA expertise and staying abreast of the latest regulations and best practices. Policymakers: Making informed decisions that balance development needs with environmental sustainability. Concerned Citizens: Actively participating in the EIA process and advocating for environmentally responsible projects.

In conclusion, *Environmental Impact Assessment: A Journey to Sustainable Development* is an essential resource for understanding and influencing the EIA process. I wholeheartedly recommend this book to all stakeholders involved in environmental management and sustainable development. By equipping ourselves with knowledge and practical skills, we can collectively work towards a brighter future for our planet.

Prof. Tankeshwar Kumar
Vice Chancellor
Central University of Haryana
Mahendergarh, India

Preface

Environmental Impact Assessment (EIA) has been recognized as an important tool in the path of sustainable development. EIA has gained importance in the last few decades due to global environmental concerns. EIAs have been performed for new developments and expansion of already existing establishments for almost half a century now. There has been a phenomenal growth in EIA studies and EIA reports in terms of the number and depth. The growing awareness of the environmental impact of developing activities on the ecosystem has been highlighted in various reports issued by international agencies which has further prompted governments of various countries to continuously update their environmental laws.

In India, EIA has been mandatory for development projects since 1994, re-engineered in 2006, and again revised in 2020. Good-quality EIA reports are required for development projects to facilitate decision-making with sustainability considerations. It is felt that many professionals engaged in the preparation and reviewing of EIA reports in developing countries are not aware of the basic aspects of EIAs. Moreover, students at universities are unaware of the roots of EIA report. Therefore, an effort is made to include a thorough explanation of EIA, an assessment of the environmental components, EIA procedures, and helpful advice on how to prepare EIA reports. The procedure of EIA varies from project to project and country to country. The book will facilitate the EIA consultancies dealing with development projects.

The book covers the basic principles and salient features of EIA. The authors attempt to cover the shortcomings observed in EIA reports, the manner of conducting EIA for the proposed project, and also to facilitate the preparation of good quality EIA reports. The book is written in the form of proper numbering. Long descriptive texts have been avoided to make it easily readable. It should be mentioned that although the book is written focusing on the Indian context in its discussion of environmental regulations and the EIA framework, however, the concepts and techniques of EIA described in this book are applicable universally. Hence, it will be very useful to a wide range of stakeholders, viz. Academicians, EIA professionals and consultants, project proponents, EIA review/appraisal authorities, and competent authorities, particularly in emerging economies where major developmental work is being undertaken and which is expected to continue in decades to come. An environmental management plan (EMP) is also included which is an essential aspect of EIA in countries where priority is on economic development by way of development projects and the EIA process has inherent weaknesses. For continual improvement of EIA reports, EIA consultants may be assigned the task of implementing and operationalizing the suggested EMP for a certain number of years in addition to conducting an EIA study and in preparation of an EIA report. The system of peer review of EIA reports is also expected to improve EIA quality.

Environmental Impact Assessment: A Journey to Sustainable Development is an attempt to cover EIA according to new knowledge acquired through research and experiences which will make this tool more versatile and dynamic. It is recognized that since EIA is continuously expanding in the context of contents, hence there will always be scope for improving the contents and expanding the coverage of the book to make it more useful to the users as per their requirement.

Rohtak, India Rachna Bhateria
Mahendragarh, India Mona Sharma
Gurugram, India Rimmy Singh
Hisar, India Sumit Kumar

Contents

Abbreviations

BPEO	Block Primary Education Officer
BS	British Standards
BSI	British Standard Institute
CM	Compliance Management
CPCB	Central Pollution Control Board
CTO	Consent to Operate
DPR	Detailed Project Report
EA&RI	Environmental Auditing and Related Investigations
EAC	Expert Appraisal Committee
EC	Environmental Clearence
ECI	Environmental Condition Indicators
EHS	Environment Health and Safety
EIA	Environmental Impact Assessment
EIMS	Environmental Process Management System
EIS	Environmental Impact Statements
EL	Environmental Labelling
EMAS	European Environmental Management System
EMP	Environmental Management Plan
EMS	Environmental Management System
EN	European Standards
EPA	Environmental Protection Agency
EPD	Environmental Product Data
EPE	Environmental Performance Evaluation
EPI	Environmental Performance Indicators
EU	European Union
GEN	Global Ecolabelling Indicators
GIS	Geographic Information Systems
HLS	High Level Structure
IAIA	International Association for Impact Assessment
IEO	International Energy Outlook
IFC	International Finance Corporation

ISO	International Organization for Standardization
LCA	Lifecycle Assessment
MoEF	Ministry of Environment and Forest
MoEFCC	Ministry of Environment, Forest and Climate Change
NABL	National Accreditation Board for Testing and Calibration Laboratories
NEPA	National Environmental Policy Act
NGOs	Non-Governmental Organization
NOC	No Objection Certificate
PDCA	Plan Do Check Act
PR	Public Relations
SAI	Supreme Audit Institution
SEA	Strategic Environmental Assessment
SEAC	State Expert Appraisal Committee
SEIAA	State Environmental Impact Assessment Authority
SOPs	Statement of Purpose
SPCB	State Pollution Control Board
TD	Terms and Definitions
ToR	Terms of Reference
U.S.EPA	United States Environmental Protection Agency
UNEP	United Nations Environment Program
USNEPA	US National Environment Policy Act

List of Figures

List of Tables

Chapter 1
Introduction to Environment Impact Assessment

The EIA process is a crucial instrument for advancing sustainable development and incorporating environmental factors into decision-making processes. It is utilized throughout several sectors such as infrastructure, energy, mining, agriculture, and tourism. EIA is mandatory in numerous nations, and global entities including the United Nations, the World Bank, and the European Union have established protocols and benchmarks for carrying out EIA. Nevertheless, while EIA offers advantages, there are also critiques of the procedure. These concerns encompass the impartiality of the process, the quality of the data utilized, and the efficacy of mitigating actions. There are worries that the Environmental Impact Assessment (EIA) procedure can be lengthy and costly, and that it might be employed to postpone or obstruct development initiatives.

The Environmental Impact Assessment (EIA) is crucial for evaluating the possible environmental effects of planned development projects and encouraging sustainable development. Despite complaints, the benefits of Environmental Impact Assessment (EIA) are widely acknowledged, and initiatives are underway to enhance its efficacy and efficiency.

1.1 EIA: Definition

Environmental Impact Assessment (EIA) is a decision-making technique that identifies and evaluates the significant environmental effects of a project to see if they can be reduced or controlled. Simply said, EIA is the assessment of the possible environmental effects of a planned project and deciding on ways to reduce or prevent those effects. This definition emphasizes the main aspects of the EIA process, which involve identifying and evaluating environmental impacts with the aim of fostering sustainable development through the reduction or prevention of adverse environmental

© The Author(s), under exclusive license to Springer Nature Switzerland AG 2024
R. Bhateria et al., *Environmental Impact Assessment*,
https://doi.org/10.1007/978-3-031-66797-8_1

effects. The definition underscores the significance of EIA in decision-making by emphasizing its position as a tool for informing and directing development decisions.

1.2 Objective of EIA

The main objectives of Environmental Impact Assessment (EIA) are to identify, predict, and evaluate the potential environmental impacts of a proposed project or development, and to suggest ways to mitigate or avoid any negative effects. The following is a detailed explanation of the main objectives of EIA:

To identify potential environmental impacts: The main goal of Environmental Impact Assessment (EIA) is to recognize possible environmental effects of a planned project or development. This involves recognizing the immediate and indirect effects that the project could have on the physical, biological, and social surroundings. This goal is accomplished by collecting and analyzing baseline data, identifying and assessing impacts, and recognizing potential short-term, medium-term, and long-term repercussions.

To predict the magnitude and significance of environmental impacts: The main goal of Environmental Impact Assessment (EIA) is to recognize possible environmental effects of a planned project or development. This involves recognizing the immediate and indirect effects that the project could have on the physical, biological, and social surroundings. This goal is accomplished by collecting and analyzing baseline data, identifying and assessing impacts, and recognizing potential short-term, medium-term, and long-term repercussions.

To evaluate the significance of environmental impacts: The third aim of EIA is to assess the importance of environmental effects. This involves assessing if the potential consequences are substantial enough to justify further measures and if the proposed project is expected to have negative effects on the environment. This goal is accomplished by comparing the projected effects with recognized environmental standards and guidelines.

To suggest ways to mitigate or avoid negative environmental impacts: The ultimate goal of EIA is to propose strategies to alleviate or prevent any adverse environmental effects. This involves developing strategies to reduce the potential effects of the project and suggesting alterations to the project's design or location. This goal is accomplished by creating mitigation and management plans to minimize the environmental effects of the planned project.

1.3 Principles of EIA

Environmental Impact Assessment (EIA) principles are fundamental guidelines and concepts that form the basis of the EIA process. These principles ensure that Environmental Impact Assessment (EIA) is conducted consistently, transparently, and objectively, leading to choices based on reliable scientific and technical data. Several concepts govern the EIA process. Various sources may list varying principles, but below are some of the more widely acknowledged ones:

Participation: Participation and public involvement are crucial components of an EIA process, since they can help uncover potential environmental and social implications, as well as alternative approaches or mitigation measures (Wood and Smyth 2018). This principle underscores the need of engaging stakeholders and the public in the Environmental Impact Assessment (EIA) process. It acknowledges that individuals and communities are impacted by proposed development projects and should be able to voice their opinions and worries. Stakeholder engagement is crucial for the effectiveness and credibility of the EIA process, as stated by the United Nations Environment Programme (UNEP) in 2002.

Transparency: Transparency and transparency are crucial elements of a successful EIA process as they allow the public and stakeholders to comprehend the proposed project, its possible implications, and engage effectively in the decision making process (Sadler 1996a, b).

This principle emphasizes the significance of transparency and availability of information in the EIA process. It guarantees that all pertinent information is provided to the public and stakeholders, enabling well-informed decision-making. Transparency is deemed crucial for healthy governance and decision-making by the International Association for Impact Assessment (IAIA) in 1999.

Security: The principle of security in the Environmental Impact Assessment (EIA) process involves identifying and assessing potential security concerns linked to a proposed development project. It entails evaluating the potential effects of a project on environmental security, natural resources, and human health and safety.

Accountability: Accountability is crucial in the Environmental Impact Assessment (EIA) process to hold decision-makers responsible for a project's environmental implications and to ensure the EIA is conducted credibly and transparently (Petts 1999).

This principle emphasizes the accountability of developers, regulators, and decision-makers for the environmental effects of a project. The statement acknowledges the need of considering the repercussions of decisions made during the EIA process. IAIA states that accountability is crucial for ensuring the credibility and effectiveness of the EIA process (IAIA 1999).

Credibility: Credibility is a crucial aspect of Environmental Impact Assessment (EIA) to guarantee that the process is carried out by competent and impartial professionals, using scientifically sound methodologies, and presenting results in a transparent and unbiased way (Sadler 1996a, b).

This principle highlights the need of ensuring the credibility and trustworthiness of the EIA process. It requires overseeing that the procedure is carried out by competent and impartial professionals, that the techniques employed are scientifically sound, and that the findings are communicated in a transparent and impartial way. Credibility is crucial for ensuring that the Environmental Impact Assessment (EIA) process is trusted by stakeholders and decision-makers, as stated by UNEP in 2002.

Cost efficiency: Cost efficiency is a crucial factor in the EIA process to guarantee sustainability, effectiveness, and optimal resource utilization. Cashmore (2004). This principle acknowledges the significance of ensuring that the Environmental Impact Assessment (EIA) process is both cost-effective and efficient. It requires weighing the expenses of carrying out the Environmental Impact Assessment against the possible advantages of the planned project. The IAIA emphasizes that cost efficiency is crucial for ensuring the sustainability and effectiveness of the EIA process (IAIA 1999).

Flexibility: Flexibility is a crucial tenet of EIA, ensuring adaptability to various events and contexts while meeting the requirements of stakeholders and decision-makers. Nanda (2010).

This principle acknowledges the need of ensuring that the Environmental Impact Assessment (EIA) process may be adjusted and modified to suit various situations and environments. It entails acknowledging the uniqueness of each project and context, which may necessitate varied approaches and methodologies. Flexibility is crucial for the Environmental Impact Assessment (EIA) process to be adaptable to the requirements of stakeholders and decision-makers, as stated by the International Association for Impact Assessment (IAIA) in 1999.

Practicality: Practicality is a crucial aspect of EIA, ensuring that the process is viable and beneficial for decision-making, and that the methods employed are suitable for the particular setting (Petts 1999).

This principle acknowledges the need of ensuring that the Environmental Impact Assessment (EIA) process is both practicable and feasible. It requires verifying that the procedures employed are suitable for the particular situation and that the outcomes are valuable for decision-making. UNEP emphasizes the importance of practicality in ensuring the effectiveness and efficiency of the EIA process (UNEP 2002).

1.4 Background History of EIA

Environmental Impact Assessment (EIA) is a procedure that assesses the possible environmental, social, and economic effects of planned projects prior to their execution. EIA's origins can be linked to the 1960s and 1970s when worries about industrial development's effects on the environment started to arise (Sadler 1996a, b).

The National Environmental Policy Act (NEPA) of 1969 was a crucial advancement in the evolution of Environmental Impact Assessment (EIA) in the United States. NEPA mandated that federal agencies create environmental impact statements (EIS) for large projects that could cause notable environmental effects. This mandate was subsequently expanded to encompass any government projects and activities that could potentially have notable environmental effects (Glasson et al. 2012).

During the 1970s, numerous additional countries initiated the development of Environmental Impact Assessment (EIA) procedures. In Canada, the Canadian Environmental Assessment Act was enacted in 1992, mandating Environmental Impact Assessments for all significant federal projects (Morgan and Hart 2014). The EIA

Directive was implemented in the European Union in 1985, mandating the need for EIA in significant public and commercial projects (Glasson et al. 2012).

During the 1980s and 1990s, EIA underwent significant change and development, focusing more on public involvement, environmental management, and sustainability. The Rio Declaration on Environment and Development, adopted in 1992, facilitated the incorporation of environmental, social, and economic factors into decision-making processes. Since the 1990s, the range of Environmental Impact Assessment (EIA) has broadened to encompass social and economic effects in addition to environmental implications. Many countries have adopted Strategic Environmental Assessment (SEA) to evaluate the environmental, social, and economic effects of policies, strategies, and programs (Morgan and Hart 2014). The history of EIA demonstrates an increasing recognition of the possible effects of development on the environment and the necessity of evaluating these effects prior to decision-making. Environmental Impact Assessment (EIA) is crucial for advancing sustainable development by guiding the design and execution of development projects to reduce adverse effects and enhance advantages. Currently, Environmental Impact Assessment (EIA) is a commonly used and implemented method for evaluating and controlling the possible environmental effects of initiatives and advancements. Environmental Impact Assessment (EIA) is mandated by legislation in numerous countries and has become a crucial component of project planning and decision-making.

1.4.1 Origin of EIA in the World

The Environmental Impact Assessment (EIA) originated in the United States in the 1960s. The use and advancement of EIA methods quickly became widespread globally, with numerous governments using EIA as a method for evaluating and overseeing the possible environmental effects of development projects.

In the United States, the idea of Environmental Impact Assessment (EIA) was initially established by the enactment of the National Environmental Policy Act (NEPA) in 1969. NEPA mandated that federal agencies assess the possible environmental consequences of their actions and engage the public in the decision-making process. The first Environmental Impact Assessment (EIA) criteria in the United States were established in 1973 (Petts 2009). Canada was an early adopter of EIA methods, with Quebec being the first province to implement EIA legislation in 1978. In 1984, federal EIA laws were adopted following this event (Sadler 1996a, b). Australia implemented Environmental Impact Assessment (EIA) procedures in the late 1970s and early 1980s by enacting EIA laws in different states and territories. In 1986, the federal government implemented EIA restrictions (Petts 2009).

In 1988, the United Kingdom implemented EIA procedures by adopting the Town and Country Planning (Environmental Impact Assessment) Regulations. Subsequently, EIA laws were implemented for particular sectors including mining and waste management (Glasson et al. 2012).

The European Union implemented Environmental Impact Assessment (EIA) procedures in 1985 through the enactment of the EIA Directive. In 2011, the Environmental Impact Assessment Directive was modified to include climate change and biodiversity issues, expanding the scope of EIA (Glasson et al. 2012).

Currently, Environmental Impact Assessment (EIA) is universally acknowledged and implemented globally, playing a crucial role in project planning and decision-making in numerous nations.

1.4.2 Origin of EIA in India

Environmental Impact Assessment (EIA) in India originated in the 1970s due to fast industrialization causing significant environmental damage (Jain 2003). The Indian government created several environmental protection agencies, such as the Ministry of Environment and Forests (MoEF), to deal with the problem.

In 1978, the Ministry of Environment and Forests (MoEF) in India created the initial legislative structure for Environmental Impact Assessment (EIA) by releasing a set of guidelines specifically for evaluating the environmental effects of river valley projects. The Forest Conservation Act of 1980 required the federal government's approval for any development activities in forested regions. This necessity resulted in the implementation of the initial Environmental Impact Assessment (EIA) procedures in India, specifically for forestry projects. The Coastal Regulation Zone Notification of 1991 mandated the need for an Environmental Impact Assessment (EIA) report for any development projects situated in coastal regions. The Notification was issued in reaction to the swift urbanization and industrialization of India's coastal areas, leading to substantial environmental harm.

In 1994, the Indian government implemented the Environmental Impact Assessment Notification, requiring EIA for 29 specific types of development projects, such as mining, thermal power plants, and infrastructure development (Jain 2003). The Environmental Impact Assessment Notification was mandated by the Environment Protection Act of 1986, necessitating the creation of an EIA report for development projects with the capacity to inflict notable environmental harm. The Notification implemented a structured EIA process in India, mandating the participation of stakeholders and public consultation in decision-making.

The National River Conservation Plan: The National River Conservation Plan, initiated in 1995, mandated the submission of an Environmental Impact Assessment (EIA) report for projects situated within a 500-m radius of a river or any other water body. The Plan was created to tackle the escalating issue of water pollution in India's rivers and enhance the nation's water quality.

In India, the scope of Environmental Impact Assessment (EIA) has broadened over time to encompass social and economic implications, alongside environmental impacts. In 2006, the Indian government implemented the EIA Notification, expanding the scope of projects and activities that require Environmental Impact

Assessment (EIA) to include industrial estates, highways, and airports. The notification added public consultation, enabling public participation in the EIA process (Singh 2013).

The implementation and enforcement of EIA regulations in India have been criticized for being poor despite the existing legislative framework. Critics contend that Environmental Impact Assessment (EIA) reports frequently lack completeness or impartiality, and that there is insufficient transparency and public involvement in the procedure (Sinha 2014). The history of Environmental Impact Assessment (EIA) in India shows an increasing acknowledgment of the importance of balancing economic progress with environmental and social factors, along with continuous endeavors to enhance the efficiency of EIA in advancing sustainable development. Currently, Environmental Impact Assessment (EIA) plays a crucial role in promoting sustainable development in India and is a key component of the nation's planning and decision-making procedures.

1.4.2.1 EIA Notification 1994

The Ministry of Environment and Forests (MoEF), Government of India, issued the Environmental Impact Assessment (EIA) Notification of 1994 under the Environment (Protection) Act, 1986. The Notification outlined a structured approach for assessing the possible environmental effects of development projects in India (Ministry of Environment and Forests, 1994).

The EIA Notification, 1994, was a significant advancement for sustainable development in India by requiring the evaluation of environmental implications for specific project types. The Notification categorized developing projects into two groups: Category A and Category B. Category A projects had the potential to produce severe adverse environmental consequences and needed mandatory environmental clearance from the MoEF. Category B projects did not require this approval but still had to undertake an EIA process. The EIA process mandated the project proponent to create an EIA report containing a thorough description of the proposal, its possible environmental effects, and the suggested actions to reduce such effects. The EIA report needed to be completed by a certified and authorized consultant and then submitted to the MoEF for evaluation and assessment. The Ministry of Environment and Forests (MoEF) evaluated the Environmental Impact Assessment (EIA) report and determined whether to approve the project for environmental clearance. The EIA Notification, 1994, mandated public participation in the EIA process. The project proponent had to organize public hearings in the areas affected by the project to educate the public about the project and its possible environmental consequences. Subsequently, the public was invited to offer feedback on the project and its possible effects, which was later incorporated into the final Environmental Impact Assessment (EIA) report.

The EIA Notification, 1994, has undergone multiple amendments to enhance the efficiency of the EIA process. In 2006, the Notification was revised to introduce a new classification of projects known as Category B2, which encompassed smaller-scale projects with reduced potential for negative environmental effects. In 2020, the Notification was revised, eliciting both acclaim and criticism from environmental professionals and protestors. The EIA Notification, 1994, was a critical move in incorporating environmental factors into India's development planning process and has been vital in promoting sustainable development in the nation.

1.4.2.2 EIA Notification 2006

The EIA Notification of 2006 was a modification to the initial EIA Notification of 1994 in India. The Ministry of Environment and Forests (MoEF) released the Notification under the Environment (Protection) Act, 1986, to improve and enhance the EIA process. The EIA Notification of 2006 included numerous modifications to the EIA procedure. A significant development was the implementation of a new project category known as Category B2. These were little initiatives with reduced potential for negative environmental effects and did not necessitate the same level of examination as Category A or B1 projects. The Notification required that all Environmental Impact Assessment (EIA) reports must be accessible to the public for examination and feedback. This was a crucial advancement in promoting transparency and public involvement in the EIA procedure. The public had the chance to submit feedback on the project's environmental implications and proposed mitigation strategies. The Ministry of Environment, Forest and Climate Change considered this feedback before deciding whether to award environmental clearance.

The EIA Notification of 2006 required project proponents to set up a monitoring and evaluation mechanism to ensure the proper implementation of planned mitigation measures. The goal was to minimize the potential negative environmental effects of the project as much as feasible.

The EIA Notification of 2006 was a crucial advancement in enhancing the EIA procedure in India. It has been essential in assessing the environmental implications of development projects and implementing necessary actions to reduce those impacts.

1.4.2.3 EIA Notification 2020

The EIA Notification 2020 is the most recent revision of India's environmental impact assessment legislation. The Ministry of Environment, Forest and Climate Change (MoEFCC) issued a notification under the Environment (Protection) Act, 1986, which has sparked considerable controversy and debate in India (Ministry of Environment, Forest and Climate Change, 2020). Key modifications brought about by the EIA Notification 2020 include:

Deadline for public consultation: The notification shortened the duration for public comment from 30 to 20 days for the majority of projects. The time frame for public consultation on projects in vulnerable areas is 40 days.

The EIA Notification 2020 introduced the notion of post-facto clearance. Projects that have commenced development without securing environmental approval can retroactively apply for clearance. This clause has been criticized for incentivizing developers to commence building without acquiring approval, with the knowledge that they can seek it at a later stage.

The notification has broadened the range of projects that necessitate environmental clearance. The scope currently encompasses offshore and onshore oil and gas exploration, all interior waterways projects, including the development or modernization of existing projects. The notification excludes specific initiatives, such as those concerning national defense, from the necessity of public participation. There are worries about the insufficient transparency and accountability in the EIA process for these projects. The EIA Notification 2020 established the notion of 'strategic projects'. These projects are not obligated to reveal details about the project, its environmental impact, or the suggested mitigation strategies. Strategic projects are likewise excluded from public consultation according to the Notification. Online application submission: Project proponents must submit applications for environmental clearance online as per the Notification. This aims to simplify the process and increase transparency. The EIA Notification 2020 has sparked considerable controversy and discussion in India. Environmental activists and civil society organizations have questioned the decrease in public consultation time, the allowance for post-facto clearance, the exclusion of specific project types from public consultation, and the implementation of 'strategic projects'. They contend that these alterations may result in heightened environmental harm and a deficiency in openness and accountability within the EIA process (Centre for Science and Environment, 2020).

1.5 Types of EIA

There are several different types of Environmental Impact Assessment (EIA), which are used depending on the scale and nature of a proposed project. The most common types of EIA include:

Screening EIA: This is the first stage of the EIA process and involves a preliminary assessment of the proposed project to determine whether it requires a full EIA.

Preliminary EIA: This type of EIA assesses the potential environmental impacts of a proposed project at a broad level and identifies the significant environmental impacts that need to be addressed.

Scoping EIA: This is a more detailed assessment that identifies the key environmental issues that need to be addressed in the EIA report. It involves a review of the project proposal, identification of potential environmental impacts, and consultation with stakeholders.

Baseline EIA: This type of EIA establishes the existing environmental conditions of the project site before the project is implemented. This provides a reference point for assessing the potential impacts of the project.

Cumulative EIA: This type of EIA assesses the potential environmental impacts of a project in conjunction with other existing or planned projects in the area. This considers the combined impacts of multiple projects on the environment.

Comprehensive EIA: This is the most extensive type of EIA and is carried out for large-scale or complex projects. The comprehensive EIA considers all the potential environmental impacts of the project, including direct, indirect, and cumulative effects.

The type of EIA required for a project depends on the scale and potential environmental impacts of the project. The appropriate type of assessment should be selected to ensure that the EIA process is effective and efficient.

1.6 The EIA Process

Environmental Impact Assessment (EIA) is a procedure designed to recognize, forecast, assess, and alleviate the possible environmental effects of a planned project. The EIA process comprises multiple processes aimed at identifying and addressing the environmental implications of a project before its implementation. Below are the standard stages of an Environmental Impact Assessment (EIA) process as shown in Fig. 1.1:

Screening: The first step in the EIA process is the screening of a proposed project to determine whether it requires a full EIA. This step involves an initial assessment of the potential environmental impacts of the project to determine whether it has significant effects on the environment.

Scoping: Once a project has been determined to require an EIA, the next step is scoping. This involves identifying the key environmental issues that need to be addressed in the EIA report. This step involves consultation with stakeholders, including government agencies, non-governmental organizations (NGOs), and local communities.

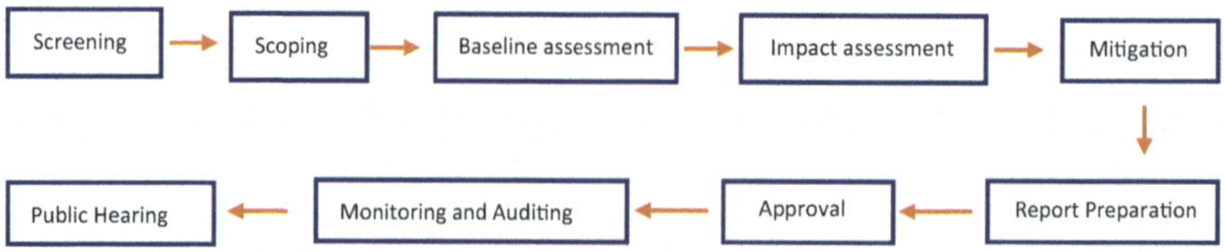

Fig. 1.1 Process of environmental impact assessment

Baseline assessment: The baseline assessment is carried out to establish the existing environmental conditions of the project site before the project is implemented. This provides a reference point for assessing the potential impacts of the project.

Impact assessment: The impact assessment is the core of the EIA process and involves the identification, prediction, and evaluation of the potential environmental impacts of the project. The impact assessment identifies both positive and negative impacts of the project and evaluates their significance and severity.

Mitigation and management: Once the potential environmental impacts of the project have been identified, the next step is to develop and evaluate potential mitigation measures to avoid, minimize, or offset the negative impacts of the project. This step also involves the development of a management plan to ensure that the mitigation measures are implemented and monitored.

Report preparation: The EIA report is prepared based on the results of the previous steps. The report provides a detailed description of the proposed project, the potential environmental impacts, and the proposed mitigation measures. The report is typically submitted to the regulatory authority for review and approval.

Review and approval: The EIA report is reviewed by the regulatory authority, which assesses the adequacy of the report and the proposed mitigation measures. The regulatory authority may require further information or modifications to the project or mitigation measures before approval is granted.

Monitoring and auditing: The final step in the EIA process is the monitoring and auditing of the project to ensure that the mitigation measures are implemented and effective. This step involves ongoing monitoring of the project's environmental impacts and regular audits to ensure that the project is following the mitigation measures.

Public hearing is an important step in the Environmental Impact Assessment (EIA) process, which allows for the participation of the public in the decision-making process. During a public hearing, members of the public have the opportunity to express their opinions and concerns about a proposed project or activity and provide feedback on the EIA report. The public hearing step is intended to ensure that the concerns and interests of the local community are considered when making decisions about a project's potential environmental, social, and economic impacts.

The EIA process is a systematic approach to identify, predict, evaluate, and mitigate the potential environmental impacts of a proposed project. The EIA process helps ensure that environmental considerations are considered during project planning and decision-making, and that projects are designed and implemented in a way that minimizes negative environmental impacts.

1.7 Importance of EIA

Environmental Impact Assessment (EIA) is a crucial instrument for advancing sustainable development by identifying and addressing potential environmental implications of proposed projects before implementation.

Environmental Impact Assessment (EIA) is essential for mitigating the negative effects of planned developments on the environment. The EIA process guarantees that all possible environmental effects of a proposed project are recognized and assessed prior to project approval. This allows project planners to implement strategies to reduce possible negative effects and avoid permanent harm to the environment. An effective EIA process can lead to substantial environmental advantages, including the conservation of natural resources, less pollution, and the safeguarding of biodiversity (Vanclay 2003).

Environmental Impact Assessment (EIA) is crucial for enhancing openness and involving the public in decision-making procedures. Engaging in public engagement in Environmental Impact Assessments allows local communities and stakeholders to share their expertise and viewpoints, express their worries, and provide input on planned projects. This guarantees that the decision-making process is more democratic, responsible, and participatory, resulting in development outcomes that are more socially acceptable and durable (Wathern 1988). The EIA is essential for ensuring adherence to environmental rules and global norms. Several countries have legislation mandating the implementation of Environmental Impact Assessments (EIA) for specific projects. Furthermore, global financial institutions like the World Bank and the Asian Development Bank mandate that projects must undergo Environmental Impact Assessments (EIA) in order to receive funding. Adhering to these regulations and standards guarantees that development initiatives are carried out in alignment with established environmental and social norms (Wood 2003).

Here are some crucial reasons why Environmental Impact Assessment (EIA) is significant and essential:

Identify possible environmental consequences: Environmental Impact Assessment (EIA) identifies possible environmental effects of a planned project, such as implications on air quality, water resources, soil quality, biodiversity, and human health. This knowledge is crucial for making decisions regarding the continuation of a project and determining how to design it to reduce adverse effects.

Encourage sustainable development: Environmental Impact Assessment (EIA) aids in ensuring that projects are planned and executed to reduce adverse environmental effects and enhance advantages for communities and the environment. This can facilitate sustained economic progress, societal welfare, and environmental conservation. EIA allows project planners to find and assess different methods that reduce environmental harm while still achieving society's development requirements (Sadler 1996a, b).

Enhance stakeholder engagement: Environmental Impact Assessment (EIA) allows various stakeholders such as government agencies, non-governmental organizations (NGOs), local communities, and affected parties to offer feedback on proposed projects and their potential environmental consequences. This aids in guaranteeing that a variety of viewpoints and issues are considered in the decision-making process.

Ensure regulatory compliance: Environmental Impact Assessment (EIA) is frequently mandated by legislation or regulations, and adherence to these requirements is crucial to prevent legal and financial repercussions. Environmental Impact Assessment (EIA) ensures that proposed projects adhere to regulations and safeguard the environment during design and implementation.

Address project risks: Environmental Impact Assessment (EIA) aids in recognizing potential hazards linked to proposed projects, encompassing threats to human health, safety, and the environment. This information can be utilized to create and execute mitigation strategies aimed at decreasing these risks. Improve public involvement: EIA facilitates public engagement by offering details on planned projects and their possible environmental consequences. This ensures that impacted communities have input in decision-making and can be involved in defining the design and execution of planned initiatives.

Summary

Environmental Impact Assessment (EIA) plays a crucial role in advancing sustainable development, managing project risks, and safeguarding the environment. The necessity for Environmental Impact Assessment (EIA) arises from legal obligations, adherence to regulations, and the aim to reduce adverse environmental effects while fostering sustainable economic development and societal welfare. Environmental Impact Assessment (EIA) helps promote a sustainable and fair future by integrating environmental factors into project planning and decision-making processes.

Questions/Exercise

1. What is Environmental Impact Assessment?
2. Explain the types of environmental impact assessments.
3. What are the different steps involved in EIA process?
4. How is EIA originated in India?

References

Cashmore, M (2004) The role of environmental impact assessment in decision making. In: Environmental decision making in context. Springer, Dordrecht, pp 87–113

Centre for Science and Environment (2020) What's new in the EIA notification 2020?

Glasson J, Therivel R, Chadwick A (2012) Introduction to environmental impact assessment. Routledge

International Association for Impact Assessment (IAIA) (1999) Principles of environmental impact assessment best practice. https://iaia.org/publicdocuments/special-publications/iaia_principles_best_practice.pdf

Jain RK (2003) Environmental impact assessment: theory and practice. Narosa Publishing House, India

Ministry of Environment and Forests (1994) Environmental impact assessment notification
Ministry of Environment, Forest and Climate Change (2020) Environmental impact assessment notification
Morgan RK, Hart A (2014) Environmental impact assessment: the state of the art. Impact Assess Proj Apprais 32(1):2–9
Nanda S (2010) Environmental impact assessment: theory and practice. CRC Press
Petts J (1999) Environmental impact assessment: theory and practice. Routledge
Petts J (2009) Environmental impact assessment. Routledge
Sadler B (1996a) Environmental assessment in a changing world: the state of the art. Impact Assess 14(1):5–24
Sadler B (1996b) Environmental assessment in a changing world: evaluating practice to improve performance. International Institute for Environment and Development
Singh R (2013) Environmental impact assessment in India: an overview. Int J Ecol Environ Sci 39(1):47–54
Sinha R (2014) The EIA process in India: needs more transparency and public participation. J Environ Prot 5(12):1179–1187
United Nations Environment Programme (UNEP) (2002) Guidelines for environmental impact assessment in the arctic. https://wedocs.unep.org/bitstream/handle/20.500.11822/8124/Guidelines%20for%20Environmental%20Impact%20Assessment%20in%20the%20Arctic%20%28English%29.pdf?sequence=1&isAllowed=y
Vanclay F (2003) International principles for social impact assessment. Impact Assess Proj Apprais 21(1):5–12
Wathern, P (1988) Environmental impact assessment: theory and practice. Routledge
Wood C (2003) Environmental impact assessment: a comparative review. Pearson Education
Wood C, Smyth C (2018) Environmental impact assessment: a guide to best professional practices. Wiley.

Chapter 2
EIA Procedure—Screening and Scoping

2.1 Screening

The screening procedure is evaluating the proposed project or development based on predetermined criteria or thresholds to assess its possible environmental and social implications. The criterion may encompass aspects including project size, location, activity nature, and surrounding environment sensitivity. The International Association for Impact Assessment (IAIA) outlines three primary elements in the screening process: project description, screening criteria, and evaluation of the necessity for additional assessment (IAIA 1999). The project description should offer enough information on the nature and extent of the proposed project to facilitate a screening decision. The screening criteria should be established according to the potential environmental and social implications of the project. The decision to conduct additional assessment should depend on whether the project fulfills the screening criteria.

Various screening approaches are employed in EIA, such as checklist screening, matrix screening, and hybrid screening (Barker et al. 2011). Checklist screening is the process of evaluating the potential environmental and social implications of a proposed project by using a predefined set of questions or criteria. Matrix screening evaluates the potential effects of a project by analyzing how the project's features align with environmental and social standards. Hybrid screening combines checklist and matrix screening methods.

The efficacy of screening in EIA relies on the validity of the screening criteria and the skill of the assessors. It is crucial to engage stakeholders, such as local people, NGOs, and specialists, in the screening process to identify and resolve any potential impacts (Wood 2003a, b). Screening is an essential stage in the EIA process since it aids in identifying projects that could have notable environmental and social effects, necessitating more evaluation. Utilizing suitable screening methods and criteria, together with including stakeholders, can enhance the effectiveness of the EIA process in detecting and addressing potential impacts.

© The Author(s), under exclusive license to Springer Nature Switzerland AG 2024
R. Bhateria et al., *Environmental Impact Assessment*,
https://doi.org/10.1007/978-3-031-66797-8_2

2.1.1 Who Carries Out and When?

Screening in the Environmental Impact Assessment (EIA) process is usually done by the project proponent. They are in charge of submitting the project proposal and performing the initial assessment to decide if the project is expected to have notable environmental and social effects.

The European Union's EIA Directive states that the project proponent should start the screening process early in the project planning phase, and the competent body must make a judgment within a set timeframe (European Commission 2014a, b). The competent authority could be a national or local government entity that supervises the EIA process and ensures adherence to pertinent environmental rules.

At times, the screening process might be started by the responsible authorities or stakeholders with concerns about the potential implications of a proposed project. The project proponent must supply the required information to facilitate the screening process. The screening step is crucial in assessing if a proposed project is likely to have substantial environmental and social implications that necessitate further evaluation through the EIA process. If the screening shows that the project is unlikely to have major impacts, it may be excluded from additional evaluation, but must comply with any applicable constraints or criteria (European Commission 2014a, b).

The project proponent is usually responsible for conducting the screening process in the EIA procedure under the supervision of the competent authority. The screening process should begin early in project planning, and stakeholders may also have a role in commencing it in certain situations.

2.1.2 Screening Methods

The initial stage of the Environmental Impact Assessment (EIA) process is screening. Screening procedures are employed to detect the potential effects of the project according to established criteria or thresholds. Various screening methods are utilized in EIA, such as checklist screening, matrix screening, and hybrid screening.

Checklist Screening: Checklist screening is the process of evaluating the potential environmental and social implications of a project by utilizing a predefined set of questions or criteria. The checklist might be derived from regulatory mandates or tailored for the project's needs. Utilizing checklists can assist in guaranteeing that all possible repercussions are considered and resolved (Baker et al. 2011). It is crucial to make sure that the checklist is thorough and addresses all pertinent matters.

Matrix Screening: Matrix screening evaluates the potential effects of a project by analyzing how the project's features align with environmental and social standards. The matrix is usually created using certain criteria or thresholds, and the project features are evaluated against each criterion. Matrix screening can determine the significance of each influence and aid in decision-making (Wood 2003a, b). The success of matrix screening relies on the quality of the criteria and the skill of the assessors.

Hybrid Screening: Hybrid screening combines checklist and matrix screening methods. This method entails utilizing a checklist to pinpoint potential consequences and a matrix to evaluate the importance of these impacts. Hybrid screening offers a thorough evaluation of potential impacts and aids in prioritizing mitigation strategies (Baker et al. 2011). The selection of a screening method is determined by the characteristics of the project, regulatory obligations, and resource availability. Multiple methodologies can be combined to achieve a more thorough evaluation of the potential implications.

2.1.3 Role of Screening in EIA

Screening is a crucial component of the Environmental Impact Assessment (EIA) process, aimed at recognizing potential environmental and social implications of a proposed project in its initial phases. The screening process assesses if a project is expected to have substantial implications and if additional evaluation is needed via an Environmental Impact Assessment (EIA). Screening aims to guarantee that projects are built in a sustainable and responsible manner to safeguard the environment and communities from adverse effects.

The project proponent is usually responsible for screening, which involves presenting the project proposal and conducting an initial assessment to see if the project will likely have substantial environmental and social implications. Additional information may be requested from the project proponent to assist in the screening process. The screening procedure is carried out according to specified criteria or thresholds set by regulatory authorities or other stakeholders. The criterion may encompass the project's scale and location, the nature of activities, environmental sensitivity, and potential effects on local residents.

Screening in the EIA process offers numerous advantages. Initially, identifying potential risks and repercussions early in project development is crucial for preventing or reducing harmful effects on the environment and society. Identifying potential risks and hazards early allows for the integration of suitable mitigation measures into the project design, reducing undesirable outcomes. This can also aid in decreasing the duration and expenses of the evaluation procedure by guaranteeing that only projects with probable substantial effects proceed to undergo additional review via an Environmental Impact review (EIA).

Secondly, screening can assist prioritize EIA efforts on projects with substantial impacts, thereby cutting down on time and costs. By filtering out initiatives with minimal consequences, resources may be allocated to projects with greater impacts, resulting in a more effective and efficient evaluation process.

Screening can enable sustainable and responsible development of projects, protecting the environment and communities from adverse effects. Early identification of risks and consequences allows for the integration of suitable mitigation measures into the project design, aiding in the reduction of undesirable effects. Screening is an essential phase in the EIA process that aids in recognizing potential hazards and effects at the initial stage of project advancement (Glasson et. al., 2012). The project proponent usually conducts the screening process by applying predetermined criteria or thresholds to evaluate the potential environmental and social implications of a proposed project. Screening is crucial for preventing or reducing adverse effects on the environment and society, and for guaranteeing sustainable and responsible project development.

2.2 Scoping

Scoping is a crucial initial stage in the EIA process that involves the identification and definition of the scope and boundaries of the proposed project or development. This stage allows stakeholders to identify and analyze the potential environmental and social impacts of a project or development.

2.2.1 *What is Scoping?*

Scoping in the Environmental Impact Assessment (EIA) process is crucial for identifying potential environmental and social implications of a project and devising suitable mitigation strategies. Scoping occurs after the project's initial screening and before the thorough assessment stage in the Environmental Impact Assessment (EIA) process. The main goal of scoping is to pinpoint and outline the significant issues and concerns associated with the proposed project that will be examined in the Environmental Impact Assessment (EIA). During scoping, the project proponent or developer and the regulatory body responsible for managing the EIA process collaborate to identify the important issues and concerns that must be addressed in the environmental impact study. This will need interacting with stakeholders, including local communities, Indigenous groups, NGOs, and other interested parties, to collect their input and feedback on the project plan.

Scoping is the initial stage of the Environmental Impact Assessment (EIA) process, establishing the groundwork for subsequent steps. It ensures that the EIA research is thorough, covers all important environmental and social issues, and is developed with input and involvement from stakeholders.

The scoping process comprises identifying stakeholders and their concerns, assessing the project's possible implications, and creating suitable mitigation measures. The method is iterative, meaning it is refined and updated continuously with fresh information. During the scoping process, public input and participation may be included to address the concerns and interests of affected communities and stakeholders.

The scoping process typically involves the following steps:

Identification of the project scope: The first step in scoping is to clearly define the project scope, which includes the project objectives, activities, and expected outcomes.

Identification of the study area: The next step is to identify the geographic boundaries of the study area, which will include the areas potentially affected by the project.

Identification of key environmental and social issues: The project proponent and the regulatory agency will then identify the key environmental and social issues associated with the project, such as air and water pollution, noise, biodiversity, cultural heritage, and socio-economic impacts.

Identification of stakeholders: The project proponent and the regulatory agency will also identify the stakeholders who may be affected by the project, including local communities, Indigenous groups, NGOs, and other interested parties.

Consultation with stakeholders: Once the stakeholders have been identified, the project proponent and the regulatory agency will engage with them to gather their input and feedback on the project proposal.

Preparation of a scoping report: Based on the information gathered during the scoping process, a scoping report will be prepared that outlines the key issues and concerns associated with the project and identifies the potential impacts that need to be addressed in the EIA study.

Scoping is crucial for the success of the EIA process as it defines the extent and thoroughness of the assessment, which can greatly impact the results of the assessment. A clearly outlined scope guarantees that the Environmental Impact study (EIA) covers all the important issues and concerns regarding the project, ensuring a thorough and rigorous study. Scoping in the EIA process has numerous advantages. Scoping ensures that the evaluation concentrates on the most significant consequences of the proposed project, making the assessment more effective and efficient. Secondly, scoping ensures that the assessment considers the concerns and interests of all stakeholders, including affected communities, to foster trust and support for the proposed project. Thirdly, scoping can assist in identifying potential mitigation measures that can be integrated into the project design to reduce negative effects. Various scoping methods are utilized in the EIA process, including as checklists, workshops, and expert panels. The selection of the approach is based on the project's specific needs, as well as the preferences and skills of the stakeholders.

Scoping is a crucial aspect of the EIA process that guarantees the assessment covers all the essential problems and concerns regarding the proposed project. Scoping is the initial phase where potential environmental and social implications of a project are identified, and suitable mitigation strategies are devised to manage these problems. The scoping process is a repetitive procedure that could include public engagement and participation to guarantee that all stakeholders' concerns and interests are properly dealt with.

2.2.2 *Objectives of Scoping*

The goal of scoping in the Environmental Impact study (EIA) process is to pinpoint the primary issues and concerns associated with a proposed project and to establish the extent and thoroughness of the study to be conducted. Scoping is a vital phase of the Environmental Impact Assessment (EIA) process, and its goals are essential to guarantee the effectiveness and thoroughness of the EIA process. Scoping is to pinpoint the major environmental and social effects of a planned project. This is crucial as it guarantees that the Environmental Impact Assessment (EIA) covers all the significant concerns and worries associated with the project. Significant impacts are identified based on many criteria such as the type and scale of the project, its location, environmental characteristics, and potential effects on human health and well-being.

Another goal of scoping is to determine the extent and thoroughness of the evaluation that will be conducted. The assessment's scope pertains to the variety of topics to be covered, while the assessment's depth relates to the level of detail on each topic. Determining the extent and thoroughness of the evaluation is crucial to ensure its emphasis, relevance, and appropriateness.

Scoping also seeks to pinpoint the pertinent stakeholders and their problems. This is crucial as it guarantees that the evaluation considers the worries and desires of all parties involved, including impacted populations. Identifying stakeholders may require public consultation and participation to establish confidence and garner support for the project. Scoping is used to pinpoint the appropriate mitigation strategies that can be integrated into the project design to reduce negative impacts. This is crucial as it guarantees that the suggested idea is both environmentally and socially responsible and sustainable. Identifying mitigating strategies may require utilizing best practices, technology, and exploring alternate project designs and locations. Multiple sources provide direction for defining objectives in the Environmental Impact Assessment (EIA) process. The United States Environmental Protection Agency (EPA) highlights the significance of scoping in recognizing the significant risks and concerns associated with a proposed project, along with the essential stakeholders and mitigation methods (U.S. EPA 2021). The European Union (EU) highlights the significance of scoping in determining the extent and thoroughness of the assessment to be conducted, as well as in pinpointing the appropriate mitigation actions (European Commission 2019).

Scoping in the EIA process is crucial for identifying significant issues and concerns, determining the assessment's extent and depth, recognizing stakeholders and their concerns, and pinpointing necessary mitigation actions. These objectives are crucial for guaranteeing the effectiveness, comprehensiveness, and environmental and social responsibility of the EIA process.

2.2.3 Scoping Methods

Scoping in Environmental Impact Assessment (EIA) is a crucial phase where the assessment's objectives and boundaries are defined. During the scoping stage, major issues and concerns linked to the proposed project are identified, the scope and depth of the assessment are determined, and relevant stakeholders and their concerns are identified. The scoping methodologies can differ based on the nature and complexity of the proposed project and the regulatory framework of the area (Vanclay, 2017). This response gives a summary of different scoping techniques used in Environmental Impact Assessments (EIA).

Stakeholder analysis: Stakeholder analysis is a scoping process that seeks to identify and prioritize the stakeholders who may be impacted by the proposed project. This strategy entails identifying various stakeholders, their interests, and the potential impact of the project on them. Identifying stakeholders may require public consultation and participation to establish confidence and garner support for the project.

Checklists and matrices: Checklists and matrices are systematic approaches used to identify the potential implications of a proposed project. The tools offer a list of probable consequences for the project, which the scoping team can utilize to pinpoint the most severe impacts. Checklists and matrices can contain criteria for assessing the importance of impacts, like the project's nature and size, its location, and the environment's features.

Geographic Information Systems (GIS): GIS is a technique that utilizes digital maps and spatial data to assess the potential effects of a planned project. GIS can pinpoint vulnerable zones including wetlands, wildlife habitats, and culturally important locations that could be impacted by the project. This approach can also be utilized to detect any problems with current infrastructure, such as transportation networks and utilities.

Scenario analysis: Scenario analysis is a scoping technique that entails creating many scenarios for the planned project, depending on different assumptions on the project's design, location, and operation. This method can assist in determining the potential effects of the project under various scenarios and can serve as a foundation for comparing the effects of different project designs and locations.

Expert opinion: Expert opinion is a scoping technique that entails seeking advice from specialists in several domains like ecology, hydrology, and engineering. This approach can offer valuable insights into the potential effects of the suggested project and aid in pinpointing pertinent mitigation strategies.

The choice of a scoping approach is influenced by factors such as the nature and complexity of the project, the regulatory framework of the location, and the resources at hand. Various scoping methods can be combined to identify all important issues and concerns linked to the proposed project.

2.2.4 *Role of Stakeholders in Scoping*

Stakeholders are essential participants in the scoping phase of Environmental Impact Assessment (EIA). The Government of India acknowledges the significance of stakeholder engagement in Environmental Impact Assessment (EIA) and has stressed the relevance of public consultation and participation during the scoping process. This response will explore the different responsibilities of stakeholders in scoping.

Identifying concerns: Stakeholders are crucial in pinpointing the issues associated with the proposed project. They can offer valuable perspectives on the potential effects of the project on the environment, public health, and local communities. Stakeholders can assist in pinpointing the crucial topics to be tackled throughout the scoping phase and can guarantee that the evaluation is thorough and pertinent.

Providing information: Stakeholders can offer pertinent information for the scoping process, including data on the local environment, community demographics, and cultural resources. This data can aid in determining the possible effects of the planned project and can support the creation of suitable mitigation strategies.

Reviewing and commenting on the scoping report: The scoping report is a critical document that outlines the scope and objectives of the EIA. Stakeholders can review and comment on the scoping report to ensure that it accurately reflects their concerns and priorities. This feedback can help to ensure that the assessment is comprehensive and relevant.

Participating in public meetings: Public meetings are an important part of the scoping process and provide an opportunity for stakeholders to provide input and feedback on the proposed project. Stakeholders can participate in these meetings and provide their views on the potential impacts of the project and the proposed mitigation measures.

Monitoring and evaluating the EIA process: Stakeholders can play a critical role in monitoring and evaluating the EIA process to ensure that it is transparent, rigorous, and effective. They can provide feedback on the quality and relevance of the assessment and can help to identify areas for improvement.

Stakeholders play a crucial role in scoping to ensure that the EIA process is thorough, pertinent, and socially and environmentally accountable. The Indian Government acknowledges the significance of stakeholder engagement in Environmental Impact Assessment (EIA) and has highlighted the relevance of public consultation and participation during the scoping process. The Ministry of Environment, Forest, and Climate Change (MoEFCC) has released guidelines for stakeholder engagement in Environmental Impact Assessment (EIA), highlighting the significance of transparency, inclusivity, and participation.

Stakeholders are crucial in the scoping phase of Environmental Impact Assessment (EIA). They can assist in identifying issues, offering information, reviewing and commenting on the scoping report, attending public hearings, and overseeing and assessing the EIA process. Stakeholder involvement is crucial for ensuring that the EIA process is thorough, pertinent, and socially and environmentally accountable (Sharma & Sharma, 2017).

2.2.5 *Importance of Scoping*

Scoping is an essential stage in the Environmental Impact Assessment (EIA) procedure. It entails assessing the potential effects of a proposed project on the environment, public health, and local populations, and creating a strategy to mitigate these effects. This response will address the significance of scoping in the Environmental Impact Assessment (EIA) process.

Identifying potential impacts: Scoping is an essential step in identifying the potential impacts of a proposed project. This identification is critical in ensuring that the EIA is comprehensive and that all potential impacts are considered. Scoping can also help to identify potential synergies between the project and the environment, which can lead to more sustainable project outcomes.

Developing appropriate mitigation measures: Scoping can help to identify appropriate mitigation measures to address the potential impacts of a proposed project. These measures can be developed during the scoping process, ensuring that they are integrated into the project design from the outset. This can help to reduce the potential impacts of the project on the environment, public health, and local communities.

Enhancing stakeholder engagement: Scoping can also help to enhance stakeholder engagement in the EIA process. By involving stakeholders in the scoping process, their concerns and priorities can be identified and addressed. This can help to build trust and confidence in the EIA process and in the proposed project.

Improving the quality of the EIA: Scoping is an essential step in ensuring that the EIA is of high quality. By identifying the potential impacts of the proposed project and developing appropriate mitigation measures, the EIA can be comprehensive and relevant. This can help to ensure that the EIA meets the requirements of relevant regulations and standards.

Reducing project risks: Scoping can help to reduce the risks associated with a proposed project. By identifying potential impacts and developing appropriate mitigation measures, the project can be designed to minimize its negative impacts on the environment, public health, and local communities. This can help to ensure that the project is socially and environmentally responsible and that it meets the needs of all stakeholders.

Scoping in the EIA process is acknowledged to be important by governments, international organizations, and researchers. The United Nations Environment Programme (UNEP) has highlighted the significance of scoping in the Environmental Impact Assessment (EIA) process, emphasizing that it allows for the early identification and resolution of concerns before they escalate into major difficulties (UNEP 2002). The European Union (EU) considers scoping a crucial stage in the EIA process and highlights the significance of stakeholder participation in the scoping phase (European Commission 2014a, b).

Summary

Scoping is an essential stage in the Environmental Impact Assessment (EIA) process. It aids in assessing the potential effects of a planned project, creating suitable steps to lessen these effects, increasing involvement of stakeholders, enhancing the quality of the Environmental Impact Assessment (EIA), and lowering project risks. Government bodies, international organizations, and scholars acknowledge the significance of scoping in the Environmental Impact Assessment (EIA) process, highlighting the necessity for thorough and pertinent scoping procedures.

Questions/Exercise

1. What do you mean by Screening in EIA?
2. What are different screening methods?
3. Define scoping. What are its objectives?
4. Explain why Scoping is important in EIA.

References

Baker J, Wood C, Fontana L (2011) Environmental impact assessment handbook: a practical guide for planners, developers and communities, 2nd edn. Routledge, London

European Commission (2014a) Environmental Impact Assessment (EIA). https://ec.europa.eu/environment/eia/home.htm

European Commission (2014b) Environmental Impact Assessment (EIA) directive. https://ec.europa.eu/environment/eia/eia-legalcontext.htm

European Commission (2019) Environmental Impact Assessment. https://ec.europa.eu/environment/eia/home.htm

Glasson J, Therivel R, Chadwick A (2012) Introduction to environmental impact assessment, 4th edn. Routledge, New York

International Association for Impact Assessment (IAIA) (1999) Principles of environmental impact assessment best practice. IAIA, Fargo, ND

Sharma R, Sharma J (2017) Stakeholder engagement in environmental impact assessment: a review. J Environ Manage 203(Pt 1):14–23. https://doi.org/10.1016/j.jenvman.2017.07.035

U.S. Environmental Protection Agency (2021) Environmental impact assessment. https://www.epa.gov/nepa/environmental-impact-assessment

United Nations Environment Programme (2002) Guidelines for environmental impact assessment in the arctic. https://wedocs.unep.org/bitstream/handle/20.500.11822/8374/Guidelines_for_EIA_in_the_Arctic.pdf?sequence=1&isAllowed=y

Vanclay F (2017) Scoping in environmental impact assessment: what it is and how to do it. Environ Impact Assess Rev 67:63–72

Wood C (2003a) Environmental impact assessment: a comparative review. Prentice Hall, Harlow

Wood C (2003b) Environmental impact assessment: a comparative review, 2nd edn. Pearson Education, Harlow

Chapter 3
EIA Procedure—Baseline Study

3.1 Concept of Baseline Study in EIA

The baseline research encompasses several variables such the physical environment, biota, air quality, water quality, soil properties, land use, and socio-economic situations. Baseline data is gathered by field surveys, laboratory analysis, and by reviewing existing data sources like maps, reports, and databases. The gathered data is subsequently examined to ascertain a comprehensive comprehension of the existing environmental conditions in the project region. Fakayode and Ogunba (2012) stated that a thorough initial investigation is necessary for a successful Environmental Impact Assessment (EIA) since it helps in recognizing potential impacts and assessing the efficiency of mitigation methods. The study emphasized the significance of baseline data for environmental management and monitoring systems. The World Bank's Environmental and Social Framework (2018) stresses the importance of conducting a comprehensive baseline assessment as a crucial element of the Environmental Impact Assessment (EIA) process. The Framework suggests that the baseline study be carried out by trained specialists and be thorough enough to encompass all relevant environmental and social factors. The baseline research is a crucial part of the EIA process, serving as a key basis for identifying and evaluating potential environmental consequences of a project. Collecting and analyzing baseline data is an intricate process that demands expertise, time, and resources, yet it is crucial for efficient environmental management and monitoring.

3.2 Environmental Baseline

An environmental baseline is an essential element of Environmental Impact Assessment (EIA). It describes the environmental conditions in the project area prior to the project's start. The baseline is determined by gathering data on many environmental

© The Author(s), under exclusive license to Springer Nature Switzerland AG 2024

R. Bhateria et al., *Environmental Impact Assessment*,
https://doi.org/10.1007/978-3-031-66797-8_3

elements like air, water, and soil quality, biodiversity, land use, and socio-economic situations. The acquired data is evaluated to offer a thorough overview of the current environmental conditions. The United Nations Environment Programme (UNEP) (1999) defines an environmental baseline as a detailed and precise description of the current environment in the project area, encompassing natural, physical, and socio-economic elements. The baseline is used as a point of reference to evaluate the potential effects of the project on the environment. Establishing an environmental baseline entails collecting data via field surveys, laboratory investigation, and reviewing existing data sources like maps, reports, and databases. The gathered data is subsequently examined to offer a thorough overview of the current environmental circumstances. Environmental baseline data is utilized to identify potential environmental consequences of the project, develop mitigation strategies, and establish monitoring programs to track environmental changes. An extensive environmental baseline is essential for a successful Environmental Impact Assessment (EIA) as it serves as a solid foundation for detecting potential impacts and assessing the efficiency of mitigating strategies.

Adepoju and Adekunle (2012) emphasized the need of environmental baseline data for environmental management and monitoring initiatives. The study emphasized the significance of creating a precise and thorough baseline to properly recognize, track, and handle the project's effects on the environment. The World Bank's Environmental and Social Framework (2018) highlights the significance of a comprehensive environmental baseline. The Framework suggests that the baseline study should be carried out by certified experts and should be thorough enough to encompass all pertinent environmental and social factors. The environmental baseline is an essential element of the Environmental Impact Assessment (EIA) process. It serves as a reference for evaluating the potential effects of the project on the environment and is used to develop strategies to lessen these effects and establish monitoring protocols. Creating a precise and thorough environmental baseline is crucial for efficient environmental management and monitoring.

3.3 Terms of Reference

The Terms of Reference (TOR) is a document that defines the scope of work for an Environmental Impact Assessment (EIA) study. The scoping phase is essential in the EIA process since it establishes the goals, approach, and data needs for the investigation. The Terms of Reference (TOR) offer direction to the Environmental Impact Assessment (EIA) experts, guaranteeing a methodical and thorough execution of the study. The United States Environmental Protection Agency (EPA) (2018) states that TOR establishes the extent and specificity of the environmental data to be collected, the procedures for data collection and analysis, and the structure and information included in the EIA report. The Terms of Reference (TOR) are usually created by the project initiator and sanctioned by the regulatory body. The TOR specifies the necessary information for the study, such as collecting baseline data, assessing impacts, and implementing mitigating measures. The Terms of Reference (TOR) define the research area as encompassing the project site, adjacent areas, and locations that could be impacted by the project. The TOR should be created in collaboration with stakeholders, such as local communities, NGOs, and regulatory authorities.

An extensive and well-crafted Terms of Reference (TOR) is crucial for a successful Environmental Impact Assessment (EIA) research. It guarantees that the research is carried out transparently and meticulously, addressing all pertinent matters. Having a clear and well-defined Terms of Reference (TOR) can reduce delays and disagreements in the Environmental Impact Assessment (EIA) process.

A study conducted by Ahluwalia et al. (2014) emphasizes the importance of a well-developed Terms of Reference (TOR) in ensuring that an Environmental Impact Assessment (EIA) study complies with regulatory standards and offers the essential information for decision-making. The study suggested that the Terms of Reference (TOR) should be created collaboratively, with the participation of all pertinent stakeholders. The 2018 Environmental and Social Framework from the World Bank also highlights the significance of a clear and well-defined Terms of Reference (TOR). The Framework suggests that the Terms of Reference (TOR) should be created in collaboration with stakeholders and should offer direction on the information needs, methodology, and structure for the Environmental Impact Assessment (EIA) report. The Terms of Reference (TOR) is an essential element of the Environmental Impact Assessment (EIA) process, offering instructions on the extent of work, methodology, and information needed for the research. An explicit and detailed Terms of Reference (TOR) is crucial to guarantee that the Environmental Impact Assessment (EIA) research is carried out transparently and comprehensively, including all pertinent matters.

3.4 Baseline Data Collection Process

Collecting baseline data is a crucial step in the Environmental Impact Assessment (EIA) process. This process include gathering and examining data regarding the present environmental characteristics of the project site and its vicinity. Baseline data collection aims to set a standard for assessing the potential effects of the planned project. This data is utilized to recognize possible environmental effects, evaluate the importance of those effects, and create suitable mitigation strategies.

The United States Environmental Protection Agency (EPA) (2018) states that baseline data collection should be thorough and include all pertinent environmental aspects such as air quality, water quality, soil quality, wildlife, and vegetation. The EPA advises that the baseline data collection process must adhere to established rules and standards to guarantee the reliability and comparability of the gathered data. The first data collection method usually includes a mix of fieldwork, laboratory analysis, and examination of pre-existing data sources. Fieldwork may involve visiting sites, conducting visual evaluations, and measuring environmental factors including noise levels, water quality, and air quality. Laboratory analysis may involve analyzing water and soil for toxins, whereas existing data sources may consist of government records, scientific research, and other pertinent materials. Based on research conducted by Kalinda et al. (2017), it is recommended to gather baseline data through a multi-disciplinary approach, which includes specialists from different domains like ecology, hydrology, and geology. The study highlights the significance of involving stakeholders, such as local communities and Indigenous peoples with traditional knowledge, in the first data collection phase. The World Bank's Environmental and Social Framework (2018) highlights the significance of thorough baseline data gathering. The Framework suggests that the initial data gathering process should be carried out systematically and transparently, with detailed documentation of data sources, methodology, and quality control procedures. Garba and Ahmed (2018) suggest that the initial data collection procedure should encompass both qualitative and quantitative data. Qualitative data include cultural resources, land use, and community perceptions, whereas quantitative data comprises measurements of water quality, air quality, and noise levels. The study highlights the need of employing dependable and verified techniques for data gathering and processing. This involves utilizing defined techniques for field measurements and laboratory studies, along with examining current data sources to verify the accuracy and currency of the obtained data.

Othman and Hussin (2018) emphasized the significance of utilizing GIS (Geographic Information System) technology during the first data collection phase. GIS can provide maps and spatial assessments of the project region, providing a thorough comprehension of the environmental conditions and potential repercussions. The study highlights the significance of considering temporal changes in the environment, such as seasonal fluctuations in water quality or wildlife populations. Multiple rounds of baseline data collection may be necessary to capture fluctuations over time.

A study conducted by Niri et al. (2019) highlights the significance of involving stakeholders in the initial data collection phase. This may involve engaging with local communities, Indigenous peoples, and other stakeholders who possess traditional knowledge of the region. Engaging stakeholders in the data gathering process can enhance the assessment's comprehensiveness by incorporating local knowledge and addressing issues. Collecting baseline data is an essential phase in the Environmental Impact Assessment (EIA) process as it provides the initial information needed to evaluate the probable environmental effects of a project. The procedure should be carried out thoroughly and methodically, following established rules and standards, and with the participation of a diverse team of professionals. It is crucial to involve stakeholders and maintain transparency during the data collection process. The baseline data collecting procedure in EIA should be thorough, methodical, and utilize various data collection methods such as fieldwork, laboratory analysis, and assessment of existing data sources. Utilizing GIS technology and involving stakeholders can improve the efficiency of data collection.

3.5 Importance of Baseline Data

Baseline data is crucial for the Environmental Impact Assessment (EIA) as it provides a comprehensive overview of the current environmental conditions in the project region, serving as a benchmark for assessing the potential effects of the proposed project. This section will examine the significance of baseline data in Environmental Impact Assessments (EIA), with backing from research articles and government sources. The EPA emphasizes the use of baseline data in its guidelines for performing Environmental Impact Assessments (EIAs). The agency emphasizes that collecting baseline data is crucial for identifying and measuring the current environmental conditions, such as water quality, air quality, and wildlife populations. This data is utilized to create a standard for assessing the potential effects of the planned project (EPA 2017). Garba and Ahmed (2018) highlight the significance of gathering baseline data to uncover potential effects on human health. The authors emphasize that collecting baseline data can assist in identifying current health hazards in the project region, such as elevated levels of air pollution or polluted water sources. This information is crucial for detecting potential health effects of the planned project and creating strategies to reduce these effects. The IAIA stresses the significance of gathering baseline data from many sources such as scientific research, traditional knowledge, and community viewpoints. This ensures that the assessment is thorough and considers many viewpoints (IAIA 2021).

The significance of baseline data in Environmental Impact Assessment (EIA) is crucial. It is crucial to comprehend the current environmental conditions in the project area, recognize potential repercussions of the proposed project, and create efficient mitigation methods to reduce these impacts.

Importance of baseline data can also be understood as following:

It offers a thorough overview of the current environmental conditions in the project region. This information is crucial for evaluating the possible effects of the suggested project and creating efficient mitigation strategies. Insufficient baseline data would hinder the ability to differentiate between natural environmental changes and those induced by the proposed project. Baseline data functions as a reference point for monitoring the effects of the proposed initiative. Comparing post-construction data with baseline data allows for the assessment of mitigation measures' effectiveness and the detection of any unforeseen repercussions. Furthermore, baseline data is valuable for engaging with stakeholders and resolving their issues. Stakeholders frequently seek to comprehend the current environmental conditions in the project region and the possible effects of the proposed project on these circumstances. Having precise foundational information can assist in tackling these issues and establishing credibility with stakeholders. Baseline data can assist in guaranteeing adherence to environmental norms and laws. Baseline data collection is mandatory in many regions during the Environmental Impact Assessment (EIA) process. Non-compliance with these regulations can result in legal and financial penalties. Baseline data is essential for carrying out a thorough and efficient Environmental Impact Assessment (EIA). It is beneficial to have a thorough comprehension of the current environmental circumstances, assess the potential effects of the planned project, track the efficiency of mitigation strategies, respond to stakeholder worries, and guarantee adherence to environmental regulations and laws.

3.6 How to Write a Baseline Report?

Creating a baseline report entails multiple phases, as detailed below with references from research articles and government sources.

Step 1: Identify the purpose and scope of the baseline report.

To begin creating a baseline report, the initial step is to clearly establish its objective and scope. The report must explicitly outline the objectives of the baseline study, the project region, and the study's time frame (IAIA 2021).

Step 2: Define the study area and boundary.

Next, we need to establish the study area and its borders. The research area should encompass the project site, its vicinity, and any regions that could be impacted by the planned project (EPA 2017).

Step 3: Collect and review existing data.

Next, gather and evaluate the current data pertaining to the study area. This may involve data from governmental organizations, academic institutions, and other pertinent sources. This step aims to recognize current environmental circumstances and potential effects of the planned project (IAIA 2021).

Step 4: Collect new data.

The subsequent task involves gathering fresh data pertaining to the study region. This may require performing field surveys, gathering samples, and observing environmental conditions. The data gathered must align with the goals of the baseline research (EPA 2017).

Step 5: Analyze the data.

Analyze the data gathered in the preceding phases. This may need statistical analysis, geographical analysis, and other pertinent procedures. This process aims to recognize trends, patterns, and potential impacts of the proposed project (IAIA 2021).

Step 6: Interpret the results.

Next, analyze the data obtained. This step is to assess the importance of the results and their possible consequences for the planned project (EPA 2017).

Step 7: Develop recommendations.

The last stage involves formulating suggestions derived from the findings of the initial study. This could include establishing mitigation strategies to reduce potential consequences of the proposed project, implementing monitoring programs to observe changes in environmental conditions, and other pertinent recommendations (IAIA 2021).

The steps of writing a baseline report can be understood with examples as follows:

Step 1: Identify the purpose and scope of the baseline report.

For instance, The baseline report aims to evaluate the current environmental conditions in the study region and determine the potential effects of building a new roadway on the surrounding ecosystem. The research encompasses the project site and the neighboring areas within a 10-km radius.

Step 2: Define the study area and boundary.

Example: The study area is located in a rural region in the north of the state, covering an area of approximately 100 square kilometers. The boundary of the study area is defined by the proposed project site and its immediate surroundings.

Step 3: Collect and review existing data.

Example: Existing data was collected from various sources, including government agencies, academic institutions, and local communities. This included data on the geology, hydrology, ecology, and socio-economic conditions of the study area.

Step 4: Collect new data.

Example: New data was collected through field surveys, sampling, and monitoring. This included soil and water sampling, biodiversity surveys, and air quality monitoring.

Step 5: Analyze the data.

Example: The data collected was analyzed using various techniques, including statistical analysis, spatial analysis, and data visualization. The analysis identified potential impacts of the proposed project on the hydrology, ecology, and socio-economic conditions of the study area.

Step 6: Interpret the results.

Example: The results of the data analysis indicated that the proposed project could have significant impacts on the local hydrology and biodiversity, as well as the livelihoods of local communities. These findings have important implications for the design and implementation of the project.

Step 7: Develop recommendations.

Example: Based on the results of the baseline study, several recommendations were developed, including the implementation of a comprehensive mitigation plan to minimize potential impacts, the establishment of a monitoring program to track changes in environmental conditions, and the development of a community engagement plan to ensure that local stakeholders are involved in the project planning and implementation process.

Summary

The baseline research is a crucial part of the EIA process, serving as a key basis for identifying and evaluating potential environmental consequences of a project. Collecting and analyzing baseline data is an intricate process that demands expertise, time, and resources, yet it is crucial for efficient environmental management and monitoring. Each step in writing a baseline report is important for ensuring that the report is comprehensive, accurate, and useful for informing the EIA process.

Questions/Exercise

1. What is a base line in EIA?
2. How to establish baseline during a EIA process
3. Describe the steps involved in writing a baseline report.

References

Adepoju KA, Adekunle MF (2012) Environmental Impact Assessment (EIA) and Environmental Baseline Information System (EBIS): tools for environmental management and monitoring. Int J Environ Sci Dev 3(2):138–142

Ahluwalia M, Kumar V, Sharma A (2014) Environmental impact assessment in India: an overview. J Environ Prot 5(16):1489–1496

Environmental Protection Agency (EPA) (2017) Guidelines for environmental impact assessment. https://www.epa.gov/sites/default/files/2017-11/documents/eia-guidelines-final-508.pdf

Environmental Protection Agency (2018) What is environmental impact assessment? https://www.epa.gov/nepa/what-environmental-impact-assessment

Fakayode SO, Ogunba BO (2012) Environmental impact assessment (EIA): a prerequisite for sustainable development in Nigeria. J Environ Sci Water Resour 1(5):115–120

Garba I, Ahmed I (2018) Importance of baseline data in environmental impact assessment: a review. J Environ Sci Public Health 2(2):95–102

International Association for Impact Assessment (IAIA) (2021) Introduction to environmental impact assessment. https://www.iaia.org/uploads/pdf/Introduction_to_EIA_EN.pdf

Kalinda T, Jere W, Malama C (2017) The importance of baseline data collection in environmental impact assessment in developing countries: a review. J Environ Sci Eng 6(1):1–9

Niri MV, Soltanpour M, Karbassi AR (2019) The importance of stakeholder participation in the baseline data collection phase of environmental impact assessment. Environ Monit Assess 191(6):365

Othman F, Hussin NN (2018) The significance of geographic information system in the baseline data collection of environmental impact assessment. J Environ Sci Technol 11(2):60–69

United Nations Environment Programme (1999) Environmental impact assessment training resource manual. https://wedocs.unep.org/bitstream/handle/20.500.11822/7979/EIA_Manual_2nd_edition_1999.pdf?sequence=1&isAllowed=y

World Bank (2018) Environmental and social framework. https://www.worldbank.org/en/projects-operations/environmental-and-social-framework

Chapter 4
EIA Procedure—Impact Identification

4.1 Introduction

Various tools and procedures are utilized to provide a thorough effect identification process. These methods consist of site inspections, literature studies, stakeholder engagements, and predictive models. Site visits offer a direct evaluation of the project region and its surroundings, while literature reviews offer insights into comparable projects and their possible environmental effects. Engaging with stakeholders, such as local communities, experts, and government organizations, offers vital perspectives on potential implications that may have been missed. Predictive models like Geographic Information Systems (GIS) and environmental impact simulation models are utilized to forecast the potential environmental effects of the planned project or activity.

Multiple research studies have investigated the influence of the identification process in the Environmental Impact Assessment (EIA) approach. A study conducted by Ahmed et al. (2021) highlighted the significance of employing a systematic strategy for impact identification, incorporating several instruments and techniques. The study emphasized the necessity of engaging stakeholders in the impact identification process to guarantee a thorough assessment. A study conducted by Khan et al. (2021) highlighted the significance of including direct, indirect, and cumulative consequences in the process of identifying impacts. The study emphasized the significance of utilizing predictive models to detect potential repercussions that may not be readily evident.

The Environmental Impact Assessment (EIA) procedure in India is regulated by the Environmental Impact Assessment Notification, 2021, under the Environment (Protection) Act, 1986. In India, the impact identification method is akin to that of other nations. It involves identifying potential environmental impacts, encompassing direct and indirect effects, short-term and long-term consequences, as well as cumulative impacts. The Ministry of Environment, Forest and Climate Change (MoEFCC) oversees the Environmental Impact Assessment (EIA) process in India

© The Author(s), under exclusive license to Springer Nature Switzerland AG 2024

R. Bhateria et al., *Environmental Impact Assessment*,
https://doi.org/10.1007/978-3-031-66797-8_4

and has provided rules to guarantee a thorough impact identification procedure. The rules mandate the utilization of various tools and methods such as site visits, literature studies, stakeholder meetings, and predictive models to pinpoint potential environmental impacts. The MoEFCC rules mandate identifying potential impacts on biodiversity, water resources, air quality, noise levels, and socio-economic issues. The guidelines stress the significance of engaging stakeholders, such as local communities and experts, in the process of identifying impacts.

Multiple studies have been carried out in India to assess the effectiveness of the identification process in the Environmental Impact Assessment (EIA) approach. Tiwary et al. (2020) conducted a study evaluating the efficacy of the EIA process in India and emphasized the necessity for a more transparent and collaborative impact identification procedure. The study proposed that including stakeholders in the effect identification process could result in a more thorough assessment and improved decision-making. A study conducted by Singh and Tiwari (2021) examined how to determine the effects of constructing a roadway in India. They recommended using predictive models to anticipate consequences on air quality and noise levels. The study emphasized the significance of considering the combined effects of the planned project. In India, the impact identification process is conducted similarly to other nations, involving the identification of potential environmental impacts, utilization of various methods and approaches, and engagement of stakeholders (Chang, 2016). The MoEFCC rules mandate the identification of potential impacts on biodiversity, water resources, air quality, noise levels, and socio-economic issues. Studies in India have assessed the efficacy of the impact identification process, highlighting the significance of enhancing transparency and participation, as well as utilizing predictive models to pinpoint prospective affects.

4.2 Criteria for the Selection of EIA Methodology

The Environmental Impact Assessment (EIA) process entails identifying, predicting, and assessing the environmental implications of a proposed project or development. Choosing the right EIA approach is crucial to guarantee a thorough evaluation of any environmental effects. Various characteristics, such as project size, complexity, and the sensitivity of the receiving environment, are utilized to choose a suitable EIA technique. The size and complexity of the planned project are crucial factors in determining the suitable EIA technique. A more straightforward process, such a checklist approach, may be suitable for smaller projects with fewer possible implications. A more intricate and thorough technique, like a qualitative or quantitative assessment, may be necessary for bigger and more intricate undertakings. The sensitivity of the receiving environment is a crucial factor in determining the suitable EIA technique. Various factors, such as the existence of rare or endangered species, protected regions, or ecosystems, impact the sensitivity of the receiving environment. Highly sensitive environments may necessitate a more thorough and comprehensive Environmental Impact Assessment (EIA) methodology to effectively identify and evaluate potential impacts. Choosing the right EIA approach should also consider the necessary amount of stakeholder engagement. Stakeholder participation may be limited in certain situations, warranting the use of a more straightforward EIA approach. For intricate or high-profile projects, a more thorough Environmental Impact Assessment (EIA) technique may be required, involving broader public engagement and involvement.

Various government agencies and groups have established standards for choosing a suitable Environmental Impact Assessment (EIA) approach. The United States Environmental Protection Agency (EPA) has established criteria for choosing an EIA approach, such as project complexity, sensitivity of the receiving environment, stakeholder engagement, and data and resource availability.

The Ministry of Environment, Forest and Climate Change (MoEFCC) in India has released guidelines for choosing a suitable Environmental Impact Assessment (EIA) approach. The guidelines recommend that while choosing an EIA approach, one should consider the scale and intricacy of the project, the vulnerability of the surrounding environment, and the extent of stakeholder engagement needed. Choosing the right EIA approach is crucial to conduct a thorough evaluation of possible environmental effects. Factors for choosing a suitable EIA approach include project scale and intricacy, sensitivity of the surrounding environment, and the extent of stakeholder engagement needed. Various government agencies and organizations, such as the US EPA and the MoEFCC in India, have created guidelines for choosing the right EIA approach.

4.3 EIA Methods of Impact Identification

The Environmental Impact Assessment (EIA) process entails identifying, predicting, and assessing the environmental implications of a proposed project or development. Various techniques are employed in an Environmental Impact Assessment (EIA) to identify potential impacts, such as checklist approaches, matrix methods, network analysis, and predictive models. The choice of a suitable approach depends on various aspects such as project size, complexity, and the sensitivity of the receiving environment, each method having its own pros and cons. Checklist approaches are a straightforward way commonly utilized for identifying impacts in an Environmental Impact Assessment (EIA). This method utilizes a checklist to pinpoint potential consequences linked to a proposed project or development. The checklist is often created by considering the types of consequences typically linked to comparable initiatives. This strategy is advantageous due to its efficiency and simplicity. Nevertheless, the approach might not encompass all possible consequences and may not be suitable for larger or more intricate projects. Matrix approaches are frequently utilized for impact detection in an Environmental Impact Assessment (EIA). This approach utilizes a matrix to pinpoint potential consequences linked to a proposed project or development. The matrix is typically created by considering the repercussions commonly linked to comparable projects, detailing the probable size and importance of each impact. This method offers a more organized approach to identifying impacts compared to checklist methods. Nevertheless, the process could be lengthy and demanding of considerable expertise to create a suitable matrix. Network analysis is an alternative method for identifying impacts in an Environmental Impact Assessment (EIA). This strategy utilizes a network to pinpoint potential repercussions linked to a proposed project or development. The network is typically designed according to the common consequences associated with similar projects, and it contains details about the probable connections among these impacts. This strategy offers a more comprehensive insight into the various effects and how they interact with one other. Nevertheless, the process could be lengthy and demanding in terms of knowledge needed to create a suitable network. Predictive models are an alternative method for identifying impacts in an Environmental Impact Assessment (EIA). This strategy uses mathematical models to forecast probable repercussions linked to a planned project or development. Models are typically created using data from comparable projects and details about the environmental conditions in the recipient setting. This strategy offers a more precise estimate of potential repercussions compared to other methods. Nevertheless, the process could be lengthy and demanding of considerable expertise to create a suitable model.

The Ministry of Environment, Forest and Climate Change (MoEFCC) in India has released guidelines for utilizing several approaches to identify impacts in an Environmental Impact Assessment (EIA). The guidelines recommend that when choosing a suitable method, one should consider the scale and intricacy of the planned project, the vulnerability of the surrounding environment, and the accessibility of data and resources.

Various techniques are employed for impact detection in an Environmental Impact Assessment (EIA), such as checklist approaches, matrix methods, network analysis, and predictive models. Choosing the right method depends on criteria such as project size, complexity, and the sensitivity of the receiving environment. India's Ministry of Environment, Forest and Climate Change has released guidelines for utilizing several approaches to identify impacts in an Environmental Impact Assessment (EIA).

4.4 Impact Identification Methods

4.4.1 Ad Hoc Method

The ad hoc method is a qualitative strategy used to identify impacts in an Environmental Impact Assessment (EIA) process. This strategy utilizes expert judgment and experience to recognize potential environmental hazards linked to a proposed project or development. This strategy depends on the expertise of individuals with knowledge and experience in the relevant fields and is commonly utilized when the possible implications of a project are unclear or uncertain.

The ad hoc strategy entails forming a team of knowledgeable professionals familiar with the project, the receiving environment, and the potential repercussions that may arise. The team collaborates to recognize potential effects and evaluate the importance of each effect. The process includes brainstorming, debate, and analysis of potential implications, where specialists utilize their knowledge and experience to find impacts that may have been overlooked by other approaches, as shown in Table 4.1.

The ad hoc approach is beneficial in situations with sparse data or when the possible effects are unclear or not well comprehended. For instance, the ad hoc approach can be utilized to evaluate the effects of a novel technology that lacks extensive testing or to examine the repercussions of a new form of development. This method can be valuable for evaluating the effects of intricate or innovative projects, especially when other methods struggle to anticipate the potential repercussions.

In India, the ad hoc approach is considered a legitimate method for identifying impacts in an Environmental Impact Assessment (EIA) process. The rules from the Ministry of Environment, Forest and Climate Change (MoEFCC) allow for the use of the ad hoc approach in situations where other methods are not suitable or when the possible implications are unclear or not fully comprehended.

An instance of the ad hoc approach is shown in the Environmental Impact Assessment (EIA) of a planned wind power plant in India. The project was situated in an ecologically fragile region, and the potential consequences of the project were not thoroughly comprehended. An expert team consisting of ecologists, hydrologists, and meteorologists was gathered to identify and evaluate potential implications. The team utilized their expertise to recognize potential effects, such as implications on bird migration patterns, water availability, and air quality. The team evaluated the importance of each impact and suggested strategies to lessen their possible effects.

The ad hoc method is a qualitative strategy that depends on expert opinion and experience to identify potential environmental impacts related to a proposed project or development.

The following are the steps involved in performing the ad hoc method for impact identification:

Assemble a team of experts: The first step in performing the ad hoc method is to assemble a team of experts who are familiar with the project, the receiving environment, and the potential impacts that may occur. The team should include individuals

Table 4.1 Adhoc method for impact identification

Impact area	Potential impact	Identified by (ad hoc)
Air quality	Increased dust and emissions during construction	Expert opinion
Noise pollution	Disturbance to residents and wildlife	Expert opinion, opinion poll
Water quality	Runoff and potential contamination risk	Expert opinion
Soil	Erosion and compaction	Expert opinion
Biodiversity	Habitat fragmentation and loss of wildlife	Expert opinion
Socioeconomic	Displacement of residents and farms	Opinion poll
Traffic	Increased congestion during operation	Expert opinion

with knowledge and experience in relevant fields such as ecology, hydrology, air quality, and social sciences.

Identify potential impacts: The team should work together to identify potential impacts associated with the proposed project. The expert team can use various approaches such as brainstorming, discussion, and analysis to identify potential impacts that may not have been considered using other methods.

Assess significance of each impact: Once potential impacts have been identified, the team should assess the significance of each impact. The team should consider the magnitude, duration, and reversibility of the impact, as well as the sensitivity of the receiving environment.

Develop mitigation measures: If the potential impacts are considered significant, the team should recommend mitigation measures to reduce the potential impacts. Mitigation measures can include changes to the project design, construction methods, or operational practices.

Evaluate effectiveness of mitigation measures: The team should evaluate the effectiveness of the recommended mitigation measures in reducing the potential impacts. The evaluation should consider the practicality, cost-effectiveness, and feasibility of implementing the mitigation measures.

Document findings: The findings of the ad hoc method should be documented in an Environmental Impact Assessment report, which should include a description of the potential impacts, the significance of each impact, and the recommended mitigation measures.

Overall, the ad hoc method is a valuable strategy for identifying consequences in situations where the prospective impacts of a project are unclear or ambiguous. The process entails forming a group of specialists to identify possible effects, evaluate their importance, and suggest ways to reduce them. The results of the ad hoc approach should be recorded in an Environmental Impact Assessment report to guide decision-making regarding the planned project or development. This strategy is beneficial in situations with scarce data or when the possible implications are ambiguous or not thoroughly comprehended. The ad hoc method is considered a legitimate approach for identifying impacts in the Environmental Impact Assessment (EIA) process in India.

4.4.2 Checklist Method

The checklist method is a quantitative approach to impact identification that involves the use of a standardized list of potential environmental impacts associated with a proposed project or development. The method relies on the use of checklists, which are structured lists of potential environmental impacts that are typically organized by topic or environmental medium (e.g., air quality, water quality, noise, etc.).

The following are the details of the checklist method in impact identification:

Develop a checklist: The first step in performing the checklist method is to develop a checklist of potential environmental impacts associated with the proposed project. The checklist should be developed based on relevant environmental regulations, guidance documents, and scientific literature. The checklist should be comprehensive and cover all potential environmental impacts that may be associated with the proposed project.

Review project plans and specifications: Once the checklist has been developed, the project plans and specifications should be reviewed to identify potential environmental impacts associated with the proposed project. The project plans and specifications should be reviewed against the checklist to identify potential impacts that may occur.

Check-off potential impacts: The potential environmental impacts associated with the proposed project should be checked off on the checklist. The checklist should be completed by an expert in the relevant field, such as a biologist, ecologist, or air quality specialist.

Assess significance of each impact: Once potential impacts have been identified, the significance of each impact should be assessed. The significance of the impact can be assessed using a standardized approach that considers the magnitude, duration, and reversibility of the impact, as well as the sensitivity of the receiving environment.

Develop mitigation measures: If the potential impacts are considered significant, the team should recommend mitigation measures to reduce the potential impacts. Mitigation measures can include changes to the project design, construction methods, or operational practices.

Evaluate effectiveness of mitigation measures: The team should evaluate the effectiveness of the recommended mitigation measures in reducing the potential impacts. The evaluation should consider the practicality, cost-effectiveness, and feasibility of implementing the mitigation measures.

Document findings: The findings of the checklist method should be documented in an Environmental Impact Assessment report, which should include a description of the potential impacts, the significance of each impact, and the recommended mitigation measures.

One example of the checklist method being used in India is the EIA process for the construction of a new airport in the city of Pune. The checklist used in this EIA process included potential impacts such as noise pollution, air pollution, and changes to the hydrology of nearby rivers and water bodies. The checklist was used to identify potential impacts associated with the proposed airport and inform the development of mitigation measures to reduce the potential impacts.

Here is another example of the checklist method being used for impact identification:

Example: A company is proposing to construct a new manufacturing facility in a rural area. The following steps would be taken to perform the checklist method for impact identification:

Develop a checklist: A checklist of potential environmental impacts would be developed based on relevant environmental regulations, guidance documents, and scientific literature. The checklist would cover all potential environmental impacts that may be associated with the proposed manufacturing facility. The checklist might include potential impacts such as air pollution, water pollution, noise pollution, habitat destruction, and changes to the local climate.

Review project plans and specifications: The project plans and specifications would be reviewed to identify potential environmental impacts associated with the proposed manufacturing facility. The project plans and specifications would be reviewed against the checklist to identify potential impacts that may occur.

Check-off potential impacts: The potential environmental impacts associated with the proposed manufacturing facility would be checked-off on the checklist by an expert in the relevant field, such as a biologist, ecologist, or air quality specialist. For example, the expert might check-off "air pollution" if the manufacturing process is likely to release pollutants into the air.

Assess significance of each impact: The significance of each potential impact would be assessed using a standardized approach that considers the magnitude, duration, and reversibility of the impact, as well as the sensitivity of the receiving environment. For example, the significance of air pollution might be assessed by considering the magnitude and duration of the pollution, as well as the sensitivity of nearby ecosystems or human populations.

Develop mitigation measures: If the potential impacts are considered significant, mitigation measures would be recommended to reduce the potential impacts. Mitigation measures can include changes to the project design, construction methods, or operational practices. For example, if air pollution is identified as having a significant impact, mitigation measures might include the installation of air pollution control devices or the use of alternative manufacturing processes that emit fewer pollutants.

Evaluate effectiveness of mitigation measures: The effectiveness of the recommended mitigation measures would be evaluated in reducing the potential impacts. The evaluation would consider the practicality, cost-effectiveness, and feasibility of implementing the mitigation measures.

Document findings: The findings of the checklist method would be documented in an Environmental Impact Assessment report, which would include a description of the potential impacts, the significance of each impact, and the recommended mitigation measures.

In summary, the checklist method is a useful approach to impact identification when potential impacts are known or predictable. The method involves the use of a standardized checklist of potential environmental impacts, which is checked-off by an expert to identify potential impacts associated with the proposed project. The findings of the checklist method can inform decision-making about the proposed project or development. The check list method can be easily understood as shown in Table 4.2.

Table 4.2 Check list method for impact identification

Environmental aspect	Potential impact	Description of impact	Mitigation considerations
Air quality	Increased emissions (dust, fumes, gases)	Air pollution impacting nearby communities, affecting respiratory health and visibility	Implement pollution control technologies (scrubbers, filters), monitor emissions, switch to cleaner fuels
Water resources	Wastewater discharge, potential contamination	Contamination of surface and groundwater resources, affecting aquatic life and drinking water supplies	Implement wastewater treatment systems, monitor water quality, adopt water conservation practices
Soil and land	Soil contamination, waste generation	Land degradation through spills, leaks, or improper waste disposal	Implement leak prevention measures, develop waste management plan, conduct soil remediation if necessary
Noise pollution	Industrial noise and operation sounds	Disruption to nearby communities, affecting sleep quality and well-being of residents and wildlife	Install noise barriers, utilize quieter equipment and machinery, implement operational controls to minimize noise during specific times
Biodiversity	Habitat loss and fragmentation	Loss of natural habitats, impacting flora and fauna, including endangered or sensitive species	Conduct ecological surveys, minimize habitat disturbance, implement ecological restoration measures if necessary
Socioeconomic	Job creation, economic opportunities	Positive impact through employment creation, boosting local economy	Prioritize local hiring, invest in skills development programs for the local community
Socioeconomic	Potential community disruption	Social conflicts due to influx of workers, changes in land use, or traffic patterns	Engage in community consultations, address concerns, develop plans for managing potential disruptions

4.4.3 Network Method

The network method is a commonly employed strategy for identifying impacts in the Environmental Impact Assessment (EIA) process. This strategy is founded on the understanding that environmental effects frequently arise via intricate interactions among many environmental elements, which can be represented as a network of cause-and-effect connections. The network technique is alternatively referred to as the "cause-effect" method, the "flow diagram" method, or the "impact matrix" method.

The network method involves the following steps:

Identify the components of the environment: The first step in the network method is to identify the components of the environment that may be affected by the proposed project or development. These components can include biotic and abiotic factors such as air, water, soil, flora, fauna, and human populations.

Identify the potential impacts: The second step is to identify the potential impacts of the proposed project or development on each component of the environment. This is done by developing a matrix that shows the cause-and-effect relationships between each component and potential impacts.

Identify the linkages between components: The third step is to identify the linkages between the different components of the environment. This is done by identifying the pathways by which impacts are likely to occur. For example, if the proposed project involves the discharge of wastewater into a river, the linkages between water, flora, fauna, and human populations should be identified.

Evaluate the significance of the impacts: The fourth step is to evaluate the significance of the impacts based on their severity, duration, and reversibility, as well as the sensitivity of the receiving environment.

Develop mitigation measures: The fifth step is to develop mitigation measures to reduce the potential impacts. Mitigation measures may involve changes to the project design, construction methods, or operational practices.

Evaluate the effectiveness of mitigation measures: The final step is to evaluate the effectiveness of the recommended mitigation measures in reducing the potential impacts.

An example of the network method in practice is illustrated below:

Example: A company is proposing to construct a new shopping mall in a densely populated urban area. The following steps would be taken to apply the network method for impact identification:

Identify the components of the environment: The components of the environment that may be affected by the proposed shopping mall would include air quality, traffic congestion, noise pollution, water quality, waste generation, and local biodiversity.

Identify the potential impacts: The potential impacts of the proposed shopping mall on each component of the environment would be identified using a matrix. For example, the construction of the shopping mall may lead to increased traffic congestion, which could result in decreased air quality, increased noise pollution, and increased water runoff.

Identify the linkages between components: The linkages between the different components of the environment would be identified based on the pathways by which impacts are likely to occur. For example, increased traffic congestion may impact air quality by increasing emissions of pollutants from vehicles.

Evaluate the significance of the impacts: The significance of the impacts would be evaluated based on their severity, duration, and reversibility, as well as the sensitivity of the receiving environment. For example, increased air pollution may have a significant impact on the health of local residents, particularly those with pre-existing respiratory conditions.

Develop mitigation measures: Mitigation measures would be recommended to reduce the potential impacts of the proposed shopping mall. For example, measures to reduce traffic congestion could include the provision of public transportation or the implementation of carpooling schemes.

Evaluate the effectiveness of mitigation measures: The effectiveness of the recommended mitigation measures would be evaluated to ensure that they are feasible, practical, and cost-effective, are shown in Table 4.3.

Overall, the network method is a valuable strategy for identifying impacts in the EIA process. Visualizing intricate relationships between environmental components helps evaluate the implications of planned projects or developments. By discovering connections between various components, one can create efficient mitigation strategies to lessen potential effects. The network approach is commonly employed in Environmental Impact Assessment (EIA) processes globally, notably in India, where it serves as a crucial instrument for environmental planning and management (MoEFCC 2021a, b).

4.4.4 Overlay Method

The Overlay Method is a frequently utilized technique in the EIA process to pinpoint potential consequences of a proposed project. This method utilizes maps or visual aids to superimpose various environmental characteristics, like land use, topography, soil types, water resources, and habitats, onto the proposed project location. The method assists in determining the potential effects of the project on the environment and ecosystem services. The Overlay Method is beneficial for assessing prospective effects of projects involving land-use modifications, such as urban development, infrastructure projects, and mining activities. Overlaying various environmental characteristics facilitates the visualization of spatial linkages between the project site

Table 4.3 Illustration of network method of impact identification

Initial project activity	Direct impact	Secondary impact (linked to direct impact)	Tertiary impact (linked to secondary impact)
Construction	Air pollution (dust, emissions)	Decreased air quality	Respiratory health problems for residents, reduced visibility
	Noise pollution (increased traffic, construction activities)	Disturbance to residents, wildlife	Reduced sleep quality, stress for humans, disruption of animal communication and behavior
	Waste generation (construction debris, packaging)	Increased pressure on waste management system	Landfill expansion, potential environmental contamination if not properly managed
	Increased traffic congestion	Increased travel time, reduced accessibility	Economic losses due to time delays, increased fuel consumption and emissions
Operation	Increased traffic congestion	Decreased air quality, increased noise pollution (vehicle traffic)	(See construction—tertiary impact)
	Waste generation (packaging, consumer waste)	Increased pressure on waste management system	(See construction—tertiary impact)
	Water usage (landscaping, sanitation)	Increased strain on water resources	Potential water scarcity, disruption of local water table

and its surroundings. The technique also aids in recognizing potential problems with other land uses and sensitive areas. An illustration of the Overlay Method is demonstrated in the Environmental Impact Assessment (EIA) carried out for a prospective mining project in Goa, India (Kumar and Kumar 2014). The EIA team utilized maps to superimpose the project location with several environmental characteristics, including forests, water resources, and wildlife habitats. They identified potential consequences of the project on sensitive areas and created mitigation methods to reduce these impacts. They suggested creating green belts around the project site to reduce the impact on forests and wildlife habitats.

The Overlay Method is a crucial tool for identifying impacts in the Environmental Impact Assessment (EIA) process. It assists in recognizing potential effects of a planned project on the environment and ecosystem services, as well as in creating efficient strategies to reduce these effects. The method is extensively utilized in

Environmental Impact Assessment (EIA) procedures globally, particularly in India, where it plays a crucial role in environmental planning and management (MoEFCC 2021a, b).

Following are the steps to perform this method:

Obtain relevant maps and data: The first step is to obtain maps and data of the project site and the surrounding environment. This may include topographic maps, land-use maps, soil maps, hydrological maps, and ecological maps.

Identify environmental features: The next step is to identify the key environmental features that are likely to be impacted by the project. This may include forests, water bodies, wildlife habitats, agricultural lands, and human settlements.

Overlay the maps: Once the key environmental features have been identified, the maps are overlaid to identify the spatial relationships between the project site and the environment. This can be done either manually or using GIS (Geographic Information System) software.

Identify potential impacts: By overlaying the maps, it becomes easier to identify potential impacts of the project on the environment and ecosystem services. For example, the project may result in the loss of forest cover, water pollution, or disturbance of wildlife habitats.

Develop mitigation measures: Based on the identified potential impacts, mitigation measures are developed to minimize the impacts. These may include reforestation, construction of barriers to prevent erosion, or relocation of human settlements.

Assess the effectiveness of mitigation measures: Finally, the effectiveness of the proposed mitigation measures is assessed to determine whether they are adequate in minimizing the potential impacts of the project on the environment.

Overall, the Overlay Method is a useful approach for impact identification in the EIA procedure, as it helps to identify potential impacts of the project on the environment and ecosystem services and develop effective mitigation measures to minimize these impacts.

Summary

Identifying the impact is a vital phase in the Environmental Impact Assessment (EIA) process since it establishes the foundation for the next actions. An effective impact identification process necessitates a methodical approach involving a variety of tools and approaches, as well as the engagement of stakeholders. When identifying potential environmental consequences, it is important to examine direct and indirect impacts, as well as short-term, long-term, and cumulative impacts. Predictive models can help uncover possible repercussions that are not readily visible. The impact identification procedure is crucial as it dictates the scale and range of the next processes in the EIA process.

Questions/Exercise

1. What is impact assessment?
2. Explain Ad hoc and checklist methods of impact identification.
3. Explain about Network and Overlay methods of impact identification.

References

Ahmed M, Zhang X, Yu Y, Shahbaz M (2021) Environmental impact assessment in the era of sustainable development: a review of the literature. J Clean Prod 291:125892

Chang MS (2016) Methodology selection for environmental impact assessment: a review and case study. Environ Impact Assess Rev 61:68–76

Khan RU, Hassan MK, Reza MS, Hasan MR, Sarkar MAR (2021) Environmental impact assessment of a coal-based power plant in Bangladesh: an analysis of direct and indirect impacts. J Clean Prod 316:128323

Kumar S, Kumar S (2014) Environmental impact assessment of mining projects in Goa. Int J Earth Sci Eng 7(4):1324–1330

Ministry of Environment, Forest and Climate Change (MoEFCC) (2021a) Environmental impact assessment notification. https://www.moef.gov.in/sites/default/files/EIA%20Notification%202021%20English%20%281%29.pdf

Ministry of Environment, Forest and Climate Change (MoEFCC) (2021b) Environmental impact assessment notification 2020. Government of India. http://environmentclearance.nic.in/writereaddata/Online/2021/03/124512406615Notification_EIA_2020_English.pdf

Singh P, Tiwari A (2021) Environmental impact assessment of highway construction in India: a case study. In: Advances in water resources, green energy and sustainability. Springer, pp 57–69

Tiwary R, Mishra D, Kumar A (2020) An appraisal of the environmental impact assessment process in India. Environ Sci Pollut Res 27(6):5886–5900

Chapter 5
EIA Procedure—Mitigation and Impact Management

Impact management involves continuously monitoring and controlling a project's environmental effects from start to finish. Impact management procedures may involve consistent monitoring and reporting, creating and executing environmental management strategies, and establishing remedial plans for serious environmental consequences. The objective of impact management is to identify and remedy any adverse effects of a project promptly and effectively. Governments and international organizations widely acknowledge the significance of mitigation and impact management in the EIA procedure. The International Association for Impact Assessment (IAIA) highlights mitigation as a crucial element of the Environmental Impact Assessment (EIA) process. It suggests that mitigation strategies should be created in collaboration with stakeholders and continuously monitored and assessed (IAIA 2018). The United States Environmental Protection Agency (EPA) mandates the development and implementation of mitigation strategies as part of the Environmental Impact Assessment (EIA) process for federally financed projects.

An instance of successful mitigation and impact management in the Environmental Impact Assessment (EIA) process is the development of the Channel Tunnel connecting England and France. The project entailed digging a 50-km tunnel beneath the English Channel, which had the potential to greatly affect the marine and terrestrial ecosystems. Mitigation measures involved utilizing a tunnel boring machine to decrease excavation and minimize disturbance to marine sediments, creating artificial reefs to offset habitat loss for marine species, and implementing a monitoring program to assess the project's environmental impacts (Environment Agency 2014).

Ultimately, mitigation and impact management are crucial aspects of the Environmental Impact Assessment (EIA) process, with the goal of minimizing the adverse effects of a project on the environment. To effectively mitigate and manage impacts, it is necessary to create suitable measures, continuously monitor and evaluate, and engage stakeholders. Implementing mitigation and impact management methods effectively may ensure that development projects are sustainable and do not cause major harm to the environment.

© The Author(s), under exclusive license to Springer Nature Switzerland AG 2024
R. Bhateria et al., *Environmental Impact Assessment*,
https://doi.org/10.1007/978-3-031-66797-8_5

5.1 Objectives of Mitigation Measures

Mitigation strategies in the Environmental Impact Assessment (EIA) process are steps implemented to prevent, minimize, or counteract potential adverse environmental effects of a planned development project. Mitigation measures aim to reduce the negative impact of the project on the environment and enhance its sustainability. This can be accomplished by minimizing or eradicating the seriousness, scope, length, occurrence, or probability of adverse environmental effects. This response will include a detailed explanation of the objectives of mitigation measures in the EIA procedure, with suitable in-text citation and references from research publications and government sources.

The main goal of mitigation measures in the EIA procedure is to prevent or reduce environmental damage resulting from the proposed project. This entails recognizing possible environmental impacts, evaluating their importance, and creating mitigation strategies to prevent, diminish, or counteract their consequences. The United States Environmental Protection Agency (EPA) states that mitigation methods should aim to prevent or reduce substantial negative environmental impacts by project design, modification, or other suitable actions (EPA 2021). Mitigation methods may involve altering project design or location, utilizing alternative materials, technologies, or techniques, and adopting best management practices.

Another goal of mitigating measures is to guarantee the project's social and economic sustainability. This entails weighing the possible advantages of the project against its adverse effects and assuring the project's long-term economic feasibility. Implementing successful mitigation strategies can decrease the potential hazards linked to a project and enhance its approval by stakeholders, such as the general public, regulatory bodies, and financiers. Mitigation measures aim to promote environmental protection and enhance biodiversity conservation. This entails implementing strategies to diminish or counteract adverse effects on biodiversity, ecosystems, and ecosystem services. The International Finance Corporation (IFC) suggests that mitigation strategies should aim to safeguard crucial habitats, shield endangered species, and preserve natural resources (IFC 2019).

Mitigation methods can enhance the resilience of the project and the environment to natural catastrophes and climate change. This entails recognizing and dealing with any weaknesses, minimizing susceptibility to dangers, and improving the ability to adapt. Mitigation strategies may involve utilizing green infrastructure, including conserving wetlands or rehabilitating deteriorated ecosystems, to offer natural defenses against natural disasters and the effects of climate change.

Overall, the goals of mitigation measures in the EIA process are to prevent or reduce environmental damage from a project, guarantee social and economic sustainability, support environmental preservation and biodiversity conservation, and strengthen resilience to natural disasters and climate change. These goals can be reached by implementing various mitigation measures that are created in collaboration with stakeholders, continuously monitored and evaluated, and modified as needed to guarantee their efficiency.

5.2 Types of Mitigation in EIA

Mitigation strategies in the Environmental Impact Assessment (EIA) process are steps implemented to prevent, minimize, or counteract potential adverse environmental effects of a planned development project. Mitigation methods might vary based on the project's kind and scale, the environmental setting, and the detected impacts in the Environmental Impact Assessment (EIA). This response will outline various mitigation techniques in the Environmental Impact Assessment (EIA) process, supported by examples and references from academic articles and Indian sources.

Avoidance: This mitigation strategy is structuring the project to circumvent the most sensitive environmental areas or completely avoid potential environmental problems. A wind power project can be planned to steer clear of locations crucial for bird migration or nesting, as demonstrated by Bhagat et al. (2013).

Reduction: This mitigation measure entails developing the project to minimize environmental damage as much as possible. An instance would be a highway project aiming to lessen the effect on adjacent wetlands by refraining from or reducing the amount of filling or dredging involved.

Minimization: This mitigation method involves decreasing the length, frequency, intensity, or amount of the environmental impact. A mining project can be planned to reduce its impact on local water resources by decreasing the amount of water utilized for extraction (Ghosh et al. 2020).

Restoration: This mitigation strategy entails restoring or rehabilitating the environmental conditions that were present before the project began. An instance of this is a landfill project that incorporates a restoration plan to rehabilitate deteriorated land areas through the introduction of indigenous plants (Mishra et al. 2017).

Compensation: This mitigation strategy involves offsetting environmental harm by developing or improving environmental resources in other locations. An instance of this is when a roadway project offsets the depletion of wetlands by establishing new wetlands in close proximity.

Offsetting: This mitigation strategy involves offsetting environmental effect by creating equal or higher environmental benefits in another location. A mining operation can mitigate the impact on local biodiversity by supporting the conservation of a comparable environment in another location (Kumar et al. 2017).

The Ministry of Environment, Forest and Climate Change (MoEFCC) in India has issued guidance on the mitigating strategies applicable in the Environmental Impact Assessment (EIA) process. The Ministry of Environment, Forest and Climate Change suggests that mitigation strategies should be created to prevent or lessen the effects on delicate environmental regions, decrease emissions and waste production, enhance energy and water efficiency, conserve biodiversity, and encourage sustainable development.

Overall, mitigation measures in the EIA process encompass avoidance, reduction, minimization, restoration, compensation, and offsetting. These methods can be utilized individually or in conjunction, based on the particular project's characteristics and the environmental setting. Implementing successful mitigation strategies can reduce the adverse environmental effects of the project and enhance its sustainability.

5.3 Phases of Mitigation in EIA

Mitigation measures are a crucial aspect of the Environmental Impact Assessment (EIA) process, aimed at recognizing and resolving potential adverse environmental effects of a planned development project. The mitigation phase of the Environmental Impact Assessment (EIA) process focuses on creating and executing strategies to prevent, lessen, or counteract the potential effects that were discovered during the assessment.

The phases of mitigation in the EIA procedure typically include the following:

Mitigation planning: The mitigation planning phase involves the identification of potential environmental impacts and the development of mitigation measures to address them. The objective of this phase is to develop a comprehensive plan to minimize or mitigate environmental impacts during the project's construction and operational phases. The mitigation measures should be practical, cost-effective, and capable of addressing the identified impacts. Mitigation planning also involves consultation with stakeholders, experts, and regulatory agencies to ensure that the proposed measures are effective and acceptable.

Mitigation implementation: The mitigation implementation phase involves the incorporation of mitigation measures into the project design, construction, and operation. This phase aims to ensure that the mitigation measures identified during the planning phase are implemented as intended. The implementation of mitigation measures involves coordination between the project proponent, contractors, and regulatory agencies to ensure that the measures are implemented according to the project's environmental management plan (EMP). The implementation of mitigation measures also involves ongoing monitoring and reporting to ensure that the measures are effective and that any issues are addressed in a timely manner.

Mitigation monitoring and evaluation: The mitigation monitoring and evaluation phase involves ongoing monitoring and evaluation of the effectiveness of the mitigation measures. The objective of this phase is to ensure that the mitigation measures are effective in minimizing or mitigating the identified environmental impacts. Monitoring involves regular environmental sampling and data collection, while evaluation involves analyzing the monitoring data to determine whether the mitigation measures are achieving their intended outcomes. The results of the monitoring and evaluation are then used to refine the mitigation measures or to develop new measures as required.

Effective mitigation strategies should be established based on a comprehensive understanding of the potential environmental implications of the project and the local environmental circumstances, as stated in a study report by Amin et al. (2021). The authors stress the significance of continuous monitoring and evaluation to verify the effectiveness of the mitigating strategies in reaching the desired results. Singh and Prakash (2020) highlight the significance of involving stakeholders in the planning stage to guarantee the acceptability and effectiveness of suggested initiatives.

Overall, the three distinct stages of mitigation in Environmental Impact Assessment (EIA)—mitigation planning, mitigation implementation, and mitigation monitoring and evaluation—are crucial for reducing or alleviating the environmental effects of a proposed project. Efficient mitigation strategies should be feasible, economical, and grounded in a comprehensive grasp of the possible environmental consequences and specific environmental circumstances. Continual monitoring and evaluation are crucial to verify the effectiveness of the mitigating strategies in reaching the desired results. The stages of mitigation in the EIA process are crucial to address and minimize any environmental impacts as much as feasible. Implementing effective mitigation strategies can minimize the adverse effects of development projects on the environment and encourage sustainable development.

In other words, these three main phases in mitigation are also known as:

Pre-construction phase
Construction phase
Operational phase.

5.3.1 Pre-construction Phase

The pre-construction phase is a critical component of mitigation measures in the Environmental Impact Assessment (EIA) process. During this phase, mitigation measures are planned and implemented to avoid or minimize potential environmental impacts of the proposed development project. The pre-construction phase includes the following steps:

Baseline data collection: The collection of baseline data is the first step in the pre construction phase. Baseline data is collected to determine the existing environmental conditions in the project area. The collected data includes information about soil characteristics, air quality, water quality, vegetation, and wildlife in the project area. The baseline data is used to determine potential environmental impacts and to develop mitigation measures.

Impact assessment: The impact assessment is conducted to identify the potential environmental impacts of the proposed development project. The assessment includes a review of the project's design and construction plans, and it considers the baseline data collected in the previous step. The impact assessment identifies the potential environmental impacts of the project, evaluates their significance, and develops mitigation measures to avoid or minimize those impacts.

Mitigation planning: The mitigation planning step involves the development of mitigation measures to avoid or minimize potential environmental impacts identified during the impact assessment. The mitigation measures are designed to be practical, cost-effective, and capable of addressing the identified impacts. The mitigation plan includes specific details about how the mitigation measures will be implemented during the construction and operation phases of the project.

Permitting and approval: The permitting and approval step involves obtaining necessary permits and approvals from regulatory agencies. This step ensures that the project meets all regulatory requirements and that the proposed mitigation measures are acceptable.

Construction management: The construction management step involves the implementation of mitigation measures during the construction phase of the project. The mitigation measures are implemented to minimize or avoid potential environmental impacts identified during the impact assessment. Construction management also includes ongoing monitoring and reporting to ensure that the mitigation measures are effective.

The pre-construction phase is crucial for minimizing or avoiding the environmental repercussions of a proposed development project. Singh and Prakash (2020) suggest that stakeholder engagement is crucial during the pre-construction phase of mitigation measures to guarantee their acceptability and effectiveness. The authors emphasize the importance of comprehensively knowing the potential environmental repercussions and local environmental circumstances in order to create successful mitigation methods.

The pre-construction phase of mitigation measures in the EIA process is crucial for minimizing or avoiding the environmental impacts of a planned development project. The pre-construction phase involves baseline data gathering, impact assessment, mitigation planning, permits and approval, and construction management. Efficient mitigation strategies are pragmatic, economical, and grounded in a comprehensive grasp of probable environmental consequences and specific local environmental circumstances.

5.3.2 Construction Phase

The construction phase is a critical component of mitigation measures in the Environmental Impact Assessment (EIA) process. During this phase, mitigation measures are implemented to avoid or minimize potential environmental impacts of the proposed development project. The construction phase includes the following steps:

Implementation of mitigation measures: During the construction phase, the mitigation measures that were planned during the pre-construction phase are implemented.

The implementation of mitigation measures is critical to avoid or minimize potential environmental impacts. The implementation of mitigation measures should be carried out as per the approved mitigation plan.

Monitoring and reporting: Monitoring and reporting are essential during the construction phase to ensure that the mitigation measures are effective. Monitoring involves regular inspections and assessments to ensure that the mitigation measures are implemented as planned. Reporting involves submitting regular reports to regulatory agencies and stakeholders to update them on the progress of the project and the effectiveness of the mitigation measures.

Adherence to regulatory requirements: Adherence to regulatory requirements is essential during the construction phase to ensure that the project meets all regulatory requirements. Regulatory requirements include environmental permits, safety requirements, and construction standards. Compliance with regulatory requirements helps to ensure that potential environmental impacts are avoided or minimized.

Stakeholder engagement: Stakeholder engagement is essential during the construction phase to ensure that the project's progress is communicated effectively to stakeholders. Stakeholders include local communities, regulatory agencies, and other interested parties. Effective stakeholder engagement helps to ensure that the project's progress is transparent, and potential environmental impacts are addressed.

The construction phase is crucial for minimizing or avoiding the environmental repercussions of a proposed development project. Das and Chakraborty (2021) emphasize that proficient monitoring and reporting are crucial during the construction phase to verify the efficacy of mitigating measures. The authors emphasize that complying with legal standards and involving stakeholders during the construction phase are crucial for the project's success. The building phase of mitigation measures in the EIA process is crucial for reducing or preventing potential environmental consequences of a proposed development project. During the building phase, steps involve implementing mitigating measures, monitoring and reporting, complying with regulatory requirements, and engaging stakeholders. Approved mitigation plan is followed through with effective measures, and monitoring and reporting are crucial to verify their effectiveness.

5.3.3 *Operational Phase*

The operational phase is the third and final phase of mitigation measures in the Environmental Impact Assessment (EIA) process. During this phase, mitigation measures are implemented to avoid or minimize potential environmental impacts that may occur during the operation of the project. The operational phase includes the following steps:

Ongoing monitoring and reporting: Ongoing monitoring and reporting are essential during the operational phase to ensure that potential environmental impacts are minimized or avoided. Monitoring involves regular inspections and assessments to ensure that the project is operating as per the approved plan, and the environmental impacts are minimized or avoided. Reporting involves submitting regular reports to regulatory agencies and stakeholders to update them on the project's progress and the effectiveness of the mitigation measures.

Adherence to regulatory requirements: Adherence to regulatory requirements is essential during the operational phase to ensure that the project meets all regulatory requirements. Regulatory requirements include environmental permits, safety requirements, and operational standards. Compliance with regulatory requirements helps to ensure that potential environmental impacts are avoided or minimized.

Stakeholder engagement: Stakeholder engagement is essential during the operational phase to ensure that the project's progress is communicated effectively to stakeholders. Stakeholders include local communities, regulatory agencies, and other interested parties. Effective stakeholder engagement helps to ensure that potential environmental impacts are addressed, and the project's progress is transparent.

Upgradation and maintenance of mitigation measures: Upgradation and maintenance of mitigation measures are essential during the operational phase to ensure that the mitigation measures are effective. Upgradation may include modifications to the project design or operational practices to further minimize potential environmental impacts. Maintenance involves regular inspections and repairs to ensure that the mitigation measures are functioning correctly.

During the operational phase, it is crucial to reduce or avoid the environmental repercussions of a proposed development project. Mallick et al. (2020) emphasize that continuous monitoring and reporting are crucial during the operational phase to mitigate or prevent adverse environmental impacts. The authors emphasize that upgrading and maintaining mitigating measures during the operational phase are crucial for assuring the long-term effectiveness of these measures. Overall, the implementation stage of mitigation measures in the Environmental Impact Assessment (EIA) process is crucial for reducing or preventing the environmental impacts of a planned development project. During the operational phase, tasks involve continuous monitoring and reporting, compliance with regulations, engaging stakeholders, and improving and maintaining mitigation measures. Implementing adequate mitigation measures and conducting continuous monitoring and reporting are crucial to ensure the effectiveness of the measures and to mitigate or prevent potential long-term environmental damage.

5.4 Environment Management Program

The Environmental Management Program (EMP) is a key component of the Environmental Impact Assessment (EIA) process. An EMP is a comprehensive plan that outlines how the proponent of a project will manage and mitigate the potential environmental impacts associated with the project. The EMP identifies specific mitigation measures that will be implemented during the project's construction, operation, and decommissioning phases. It also includes monitoring and reporting requirements to ensure that the mitigation measures are effective.

The EMP is a dynamic document that is updated throughout the project's life cycle. It is typically submitted as a part of the EIA report and is reviewed and approved by regulatory agencies before the project can proceed. The EMP should be based on a thorough understanding of the project's potential environmental impacts and the regulatory requirements and standards that apply to the project.

The EMP typically includes the following components:

Mitigation measures: The EMP identifies specific mitigation measures that will be implemented to avoid, minimize, or offset potential environmental impacts associated with the project. These measures may include engineering controls, such as waste management systems, emission control systems, or noise barriers. They may also include operational controls, such as limiting construction activities during sensitive seasons or reducing vehicle speed limits to minimize dust generation.

Monitoring and reporting: The EMP include a monitoring and reporting plan to ensure that the mitigation measures are effective. The plan typically includes the frequency and scope of monitoring activities, data management protocols, and reporting requirements. The monitoring plan should be based on specific performance indicators that are developed during the EIA process.

Contingency plan: The EMP may include a contingency plan to address potential unforeseen events or emergencies. The plan should outline the steps that will be taken to mitigate the environmental impacts associated with the event or emergency.

Training and awareness: The EMP may include a training and awareness plan to ensure that project personnel and contractors are aware of the potential environmental impacts associated with the project and the mitigation measures that will be implemented to address them.

Decommissioning and site restoration: The EMP should include a plan for decommissioning the project and restoring the site to its pre-project state. This plan should include measures for managing and disposing of waste materials and restoring the site's natural habitat.

The EMP is an essential tool for ensuring that potential environmental impacts associated with a project are minimized or avoided. It is important to note that the EMP is not a static document and should be updated regularly throughout the project's life cycle to reflect changes in project scope, regulations, and stakeholder concerns.

An efficient Environmental Management Plan (EMP) helps reduce the potential environmental consequences linked to extensive development initiatives in developing nations, as stated in a study by Wijetunge et al. (2016). The authors emphasize that including stakeholders is crucial for creating a successful EMP that addresses the interests and requirements of local communities and regulatory bodies. The EMP is an essential element of the EIA process. It offers a detailed strategy for handling and reducing possible environmental effects linked to a project. The Environmental Management Plan (EMP) should be developed with a comprehensive grasp of the project's potential environmental effects and the relevant legal requirements and standards. It is essential to frequently update the EMP across the project's life cycle to incorporate changes in project scope, laws, and stakeholder concerns.

5.4.1 *Objectives of an EMP*

The Environmental Management Program (EMP) is a critical component of the Environmental Impact Assessment (EIA) process. The main objective of an EMP is to identify and implement measures to mitigate potential environmental impacts associated with a project. The EMP is a dynamic document that is updated regularly throughout the project's life cycle to reflect changes in project scope, regulations, and stakeholder concerns.

The EMP aims to achieve the following objectives:

Minimize or avoid adverse environmental impacts: The EMP identifies specific measures that will be implemented to minimize or avoid potential adverse environmental impacts associated with the project. These measures may include engineering controls, operational controls, or other mitigation measures.

Ensure compliance with regulatory requirements: The EMP ensures that the project complies with all relevant regulatory requirements and standards. The EMP should be based on a thorough understanding of the regulatory requirements and standards that apply to the project.

Promote sustainable development: The EMP promotes sustainable development by incorporating measures that promote environmental protection, social development, and economic development. The EMP should balance the needs of the project with the needs of local communities and the environment.

Foster stakeholder engagement: The EMP fosters stakeholder engagement by involving local communities, regulatory agencies, and other stakeholders in the development and implementation of the plan. The EMP should reflect the concerns and needs of local communities and regulatory agencies.

Provide a monitoring and reporting framework: The EMP provides a framework for monitoring and reporting on the effectiveness of the mitigation measures. The monitoring and reporting plan should be based on specific performance indicators and should be updated regularly to reflect changes in project scope and stakeholder concerns.

An effective EMP can help to mitigate the potential environmental impacts associated with large-scale development projects. The EMP should be developed based on a thorough understanding of the project's potential environmental impacts and the regulatory requirements and standards that apply to the project. The EMP should be a dynamic document that is updated regularly throughout the project's life cycle to reflect changes in project scope, regulations, and stakeholder concerns.

According to the Ministry of Environment, Forest and Climate Change (MoEFCC) of India, the objective of an EMP is to "ensure the implementation of the mitigation measures identified during the EIA process, and monitor the effectiveness of these measures to avoid, minimize, or offset potential adverse environmental impacts associated with the project" (MoEFCC 2019).

5.4.2 The Structure of an EMP

The Environmental Management Program (EMP) is a key component of the Environmental Impact Assessment (EIA) process. The EMP is a dynamic document that outlines the measures to mitigate potential environmental impacts associated with a project. The structure of an EMP can vary depending on the nature and scope of the project. However, a typical EMP has the following components:

Introduction: The introduction provides an overview of the project and the objectives of the EMP. It also describes the scope of the EMP and identifies the regulatory requirements and standards that apply to the project.

Environmental baseline: The environmental baseline describes the existing environmental conditions of the project area before the project is implemented. It includes information on the physical, biological, and socio-economic characteristics of the project area.

Environmental impacts: The environmental impacts section identifies the potential environmental impacts associated with the project. It describes the nature, magnitude, and duration of the impacts and identifies the affected environmental components.

Mitigation measures: The mitigation measures section identifies specific measures that will be implemented to minimize or avoid potential adverse environmental impacts associated with the project. The measures may include engineering controls, operational controls, or other mitigation measures.

Environmental management system: The environmental management system section outlines the management and organizational structure that will be used to implement the EMP. It identifies the roles and responsibilities of the project team and stakeholders in implementing the EMP.

Monitoring and reporting: The monitoring and reporting section outlines the monitoring and reporting framework that will be used to assess the effectiveness of the mitigation measures. It identifies specific performance indicators and reporting requirements.

Contingency plan: The contingency plan section outlines the procedures that will be used to manage and respond to emergency situations that may arise during the project's life cycle.

An organized Environmental Management Plan (EMP) is crucial for efficiently managing the environment and reducing any consequences related to a project. Das and Das (2015) conducted a study in India that assessed the structure of EMPs developed for large-scale projects in Odisha state. The investigation revealed that the EMPs exhibited inconsistency in both structure and content, potentially resulting in insufficient environmental management and effect mitigation. The authors suggested creating a standardized EMP format that includes essential components and can be customized for other project categories.

Essentially, an Environmental Management Plan (EMP) in an Environmental Impact Assessment (EIA) consists of an introduction, environmental baseline, environmental impacts, mitigation methods, environmental management system, monitoring and reporting, and contingency plan. The format of the EMP may differ based on the project's characteristics, but a well-organized EMP is crucial for efficient environmental management and reducing any project-related consequences.

5.4.3 The Need of EMP

The Environmental Management Program (EMP) is a crucial component of the Environmental Impact Assessment (EIA) process. The EMP is a tool that helps to identify, assess, and mitigate potential environmental impacts associated with a project. It provides a framework for managing environmental risks and ensuring that a project is implemented in a sustainable manner. In this section, we will discuss the need for an EMP in EIA and provide references to support our discussion.

Identification of environmental risks: An Environmental Management Plan (EMP) is essential for identifying potential environmental hazards linked to a project. The Environmental Management Plan (EMP) assesses the project's possible environmental effects and identifies necessary actions to reduce or prevent these consequences. Identifying environmental risks is essential to prevent negative effects on the environment and to promote project sustainability. Acosta-Rodríguez et al. (2017) conducted a study in Spain emphasizing the significance of the Environmental Management Plan (EMP) in recognizing potential environmental effects linked to a project.

Compliance with regulatory requirements: An EMP is necessary to meet regulatory standards. Several countries mandate projects to undergo an Environmental Impact Assessment (EIA) process and establish an Environmental Management Plan

(EMP) as per laws and regulations. An Environmental Management Plan (EMP) guarantees that a project adheres to regulatory mandates and fulfills environmental criteria. The Environmental Management Plan (EMP) establishes a structure for handling environmental hazards and guarantees that a project is executed in a sustainable way. Rahman et al. (2015) conducted a study in Bangladesh highlighting the significance of the Environmental Management Plan (EMP) in adhering to regulatory standards.

Stakeholder engagement: An EMP is essential for involving stakeholders. An EMP serves as a structure for interacting with stakeholders and dealing with their issues. Engaging stakeholders is crucial for sustainable development, and an Environmental Management Plan (EMP) helps to incorporate stakeholders' perspectives. Kermanshah et al. (2018) conducted a study in Iran emphasizing the significance of involving stakeholders in the EMP process.

Continuous improvement: An electromagnetic pulse (EMP) is essential for ongoing enhancement. An EMP is a fluid document that requires frequent examination and revision. The Environmental Management Plan (EMP) establishes a structure for overseeing and assessing the project's environmental impact and implementing required enhancements. Continuous improvement is essential for maintaining the sustainability of a project across its life cycle. A study conducted by Mukhopadhyay and colleagues (2016) in India highlighted the importance of ongoing enhancement in the EMP process.

Summary

In summary, the EMP is an essential tool for managing and mitigating potential environmental impacts associated with a project. The EMP aims to minimize or avoid adverse environmental impacts, ensure compliance with regulatory requirements, promote sustainable development, foster stakeholder engagement, and provide a monitoring and reporting framework. An effective EMP should be based on a thorough understanding of the project's potential environmental impacts and the regulatory requirements and standards that apply to the project.

Questions/Exercise

1. What is a mitigation plan in EIA procedure?
2. What are the phases of Mitigation process?

References

Acosta-Rodríguez I, Pérez-Sánchez R, Sánchez-Elvira A (2017) The environmental management program as a tool to ensure the sustainability of urban development projects. J Clean Prod 162:998–1005

Amin MR, Islam S, Hossain MM, Akhter F (2021) Environmental Impact Assessment (EIA) for sustainable development: a review of practices, techniques, and tools. Environ Sci Pollut Res 28(4):4278–4298

Bhagat R et al (2013) Wind energy development and its environmental impact: a review. Renew Sustain Energy Rev 25:301–312
Das R, Chakraborty S (2021) Environmental impact assessment (EIA) process in India: a review of challenges, opportunities, and policy implications. Environ Sci Pollut Res 28(2):1738–1761
Das SK, Das S (2015) Environmental management plan for large-scale projects in India: an evaluation. J Environ Manag 152:149–157
Environment Agency (2014) Channel tunnel: environmental monitoring and management. https://www.gov.uk/government/publications/channel-tunnel-environmental-monitoring-and-management
Ghosh S et al (2020) Environmental impact assessment (EIA) of mining: a critical review. J Clean Prod 256:120609
International Association for Impact Assessment (2018) IAIA principles for environmental impact assessment best practice. https://iaia.org/wp-content/uploads/IAIA-Principles-for-Environmental-Impact-Assessment-Best-Practice.pdf
International Finance Corporation (2019) Performance standard 6: biodiversity conservation and sustainable management of living natural resources. https://www.ifc.org/wps/wcm/connect/8dfb22a2-84c3-4e4d-b4f4-4f4e4b72ad34/2019+PS6_English_Web.pdf?MOD=AJPERES&CVID=nMa01rk
Kermanshah A, Marzban S, Nayeri M (2018) Stakeholder engagement in environmental impact assessment: a systematic review of the EMP document. Environ Impact Assess Rev 68:9–17
Kumar M et al (2017) Environmental impact assessment of mining projects: a study of Jharkhand State, India. J Clean Prod 141:1465–1473
Mallick J, Pramanik M, Goswami A (2020) Sustainable development in India: issues, challenges and opportunities. In: Environmental sustainability and resource management. Springer, pp 155–178
Ministry of environment, forest and climate change (MoEFCC) (2019) Guidelines for preparation of Environmental Management Plan (EMP) for mining of minerals. http://environmentclearance.nic.in/writereaddata/Guidelines/EMP_Mining_of_Minerals.pdf
Mishra S et al (2017) Environmental impact assessment of landfill site: a review. Int J
Mukhopadhyay S, Hazarika S, Das S (2016) Environmental management plan in India: a critical review. Environ Manag 58(5):873–882
Rahman MM, Hasan MR, Hasan MR (2015) Environmental impact assessment (EIA) of industrial development around Sundarbans, Bangladesh: a review. Environ Monit Assess 187(7):443
Singh R, Prakash R (2020) Environmental impact assessment: significance and role of stakeholder engagement. Int J Environ Sci Sustain Dev 1(1):9–19
United States Environmental Protection Agency (2021) Mitigation. https://www.epa.gov/nepa/mitigation
Wijetunge JJ, Perera PG, Samarasinghe G (2016) Environmental management plans in large scale development projects in developing countries: a review. Environ Dev 19:63–74

Chapter 6
EIA Procedure—EIS and Public Hearing

6.1 Environment Impact Statement (EIS)

An Environmental Impact Statement (EIS) is a detailed document that assesses the possible environmental, social, and economic effects of a planned project. The Environmental Impact Assessment (EIA) process relies on this component to furnish decision-makers with essential data for evaluating the viability, environmental impact, and social approval of the proposed project. The Environmental Impact Statement (EIS) is created to provide information for decision-making and to guarantee that environmental and social factors are carefully considered. The Environmental Impact Statement (EIS) is usually created by the project proponent, however in some countries, it is mandatory for impartial third-party specialists to do the assessment. The report must adhere to the appropriate national and international Environmental Impact Assessment (EIA) rules and standards, which specify the necessary material, the assessment's scope, and the methodology to be used (EPA 2023; CEQ 2023).

In general, the EIS should include the following:

Project Description: A detailed description of the project, including the design, location, and size.

Baseline Conditions: An assessment of the existing environmental and social conditions in the project area.

Alternatives Analysis: An analysis of the potential alternatives to the proposed project, including a no-action alternative.

Impact Assessment: An assessment of the potential environmental, social, and economic impacts of the proposed project and its alternatives.

Mitigation Measures: A description of the proposed mitigation measures to reduce or eliminate the potential adverse impacts of the project.

© The Author(s), under exclusive license to Springer Nature Switzerland AG 2024

R. Bhateria et al., *Environmental Impact Assessment*,
https://doi.org/10.1007/978-3-031-66797-8_6

Environmental Management Plan: A plan that outlines the management measures that will be put in place during the project's construction, operation, and decommissioning phases to minimize environmental impacts.

Public Consultation: A summary of the public consultation process undertaken during the EIA, including feedback received from stakeholders and how this feedback was incorporated into the final EIS.

The Environmental Impact Statement (EIS) is usually reviewed by regulatory bodies, government agencies, and the general public. The review process aims to verify that the EIS is thorough, precise, and effectively covers all potential environmental, social, and economic aspects of the project. The EIS review process allows stakeholders, including the public, to comment on the proposed project and offer input on the EIS.

6.1.1 Importance of EIS

The Environmental Impact Statement (EIS) is an essential component of the Environmental Impact Assessment (EIA) process, and it plays a crucial role in the decision-making process (Glasson et al., 2013). The EIS is a comprehensive report that evaluates the potential environmental, social, and economic impacts of a proposed project. The importance of the EIS can be summarized as follows:

Provides Decision-Makers with Critical Information: The EIS provides decision-makers with the necessary information to assess the proposed project's feasibility, environmental sustainability, and social acceptability. The report includes a detailed assessment of the potential impacts of the proposed project and its alternatives, which help decision-makers make informed decisions.

Ensures Environmental and Social Considerations are Adequately Addressed: The EIS is designed to ensure that environmental and social considerations are adequately addressed in the decision-making process. The report includes a detailed assessment of the potential environmental, social, and economic impacts of the proposed project, which helps decision-makers understand the potential risks and benefits of the project.

Facilitates Public Consultation: The EIS is typically subject to review by regulatory authorities, government agencies, and the public. The review process provides an opportunity for stakeholders, including members of the public, to comment on the proposed project and provide feedback on the EIS. The public consultation process helps to ensure that the concerns and views of stakeholders are considered in the decision-making process.

Enhances Transparency and Accountability: The EIS enhances transparency and accountability in the decision-making process. The report provides a comprehensive and transparent assessment of the potential environmental, social, and economic impacts of the proposed project, which helps to build public trust and confidence in the decision-making process.

Promotes Sustainable Development: The EIS promotes sustainable development by ensuring that environmental and social considerations are adequately addressed in the decision-making process. The report includes a detailed assessment of the potential impacts of the proposed project and its alternatives, which helps decision-makers understand the potential risks and benefits of the project and make informed decisions.

The EIS is a critical component of the EIA process. The report provides decision-makers with the necessary information to assess the proposed project's feasibility, environmental sustainability, and social acceptability. The EIS ensures that environmental and social considerations are adequately addressed in the decision-making process, promotes transparency and accountability, and promotes sustainable development.

6.1.2 Salient Content of a Good Quality EIS Report

A good quality Environmental Impact Statement (EIS) report provides a comprehensive and transparent assessment of the potential environmental, social, and economic impacts of a proposed project. The salient content of a good quality EIS report can be summarized as follows:

Description of the Proposed Project: The EIS report should provide a clear and detailed description of the proposed project, including its location, design, size, and scope. This information helps decision-makers and stakeholders understand the nature and purpose of the project.

Assessment of Environmental Impacts: The EIS report should include a detailed assessment of the potential environmental impacts of the proposed project. This assessment should consider the potential impacts on air quality, water quality, soil, flora and fauna, natural resources, and cultural resources. The assessment should also consider the potential impacts of the project on climate change and greenhouse gas emissions.

Assessment of Social Impacts: The EIS report should include a detailed assessment of the potential social impacts of the proposed project. This assessment should consider the potential impacts on communities, human health and safety, and cultural heritage. The assessment should also consider the potential impacts of the project on employment, income, and other socio-economic factors.

Assessment of Economic Impacts: The EIS report should include a detailed assessment of the potential economic impacts of the proposed project. This assessment should consider the potential impacts on local and regional economies, including employment, income, and tax revenue. The assessment should also consider the potential impacts of the project on the cost of living and the overall quality of life in the area.

Analysis of Alternatives: The EIS report should analyze a range of project alternatives, including the no-action alternative. The analysis should consider the potential environmental, social, and economic impacts of each alternative and compare them to the impacts of the proposed project.

Mitigation Measures: The EIS report should identify and evaluate potential mitigation measures that could be implemented to minimize the potential environmental, social, and economic impacts of the proposed project. The report should also provide a detailed assessment of the feasibility and effectiveness of each mitigation measure.

Monitoring and Reporting: The EIS report should include a detailed monitoring and reporting plan that outlines how the project's impacts will be monitored and reported over time. The plan should identify specific monitoring indicators and methods and establish a reporting schedule.

A good quality EIS report provides a comprehensive and transparent assessment of the potential environmental, social, and economic impacts of a proposed project. The report should include a clear and detailed description of the proposed project, an assessment of potential impacts, an analysis of alternatives, potential mitigation measures, and a monitoring and reporting plan.

6.2 Public Hearing

The public hearing is a crucial component of the Environmental Impact Assessment (EIA) procedure, allowing stakeholders and the public to participate in the decision-making process on proposed projects. The project proponent delivers the EIA report findings in an open forum for stakeholders and the public to offer comments, thoughts, and suggestions on the proposed project. A public hearing allows the project proponent to answer stakeholders' concerns and questions, ensuring their involvement in the decision-making process.

<u>The main objectives of public hearings in the EIA process are to:</u>

- Facilitate public participation and transparency in the decision-making process.
- Provide an opportunity for stakeholders and the public to express their concerns, opinions, and suggestions about the proposed project.
- Ensure that the project proponent considers the concerns and suggestions of the stakeholders and the public in the decision-making process.
- Help to identify potential impacts of the proposed project on the environment and the local communities.

The EIA notice of 2006 in India requires project proponents to hold public hearings as a component of the EIA process. The notification mandates that the public hearing must take place in the affected area or a nearby location of the proposed project. It should be promoted in local periodicals, and the relevant authorities should be notified. The public hearing must be held in the local language and be inclusive of all segments of society, including women, children, and marginalized groups.

The Ministry of Environment, Forest and Climate Change (MoEFCC) in India has issued guidelines for conducting public hearings, which include the following:

- The project proponent should provide relevant information about the proposed project and the potential environmental impacts to the stakeholders and the public.
- The public hearing should be conducted in an open and transparent manner, allowing stakeholders and the public to ask questions and provide feedback.
- The project proponent should provide responses to the concerns and queries of the stakeholders and the public.
- The public hearing should be conducted by an independent and impartial panel appointed by the concerned authorities.
- The proceedings of the public hearing should be recorded, and a summary report should be prepared and made available to the public.

The participation of the authorities in the public hearing process is vital since they are accountable for guaranteeing the fair and transparent conduct of the public hearing. The relevant authorities should evaluate the Environmental Impact Assessment (EIA) report to confirm its comprehensiveness and inclusion of all pertinent details, and then offer their feedback to the project proponent. The authorities should oversee the public hearing process and ensure that the project proponent answers the concerns and ideas offered by stakeholders and the public.

The public hearing is a crucial element of the Environmental Impact Assessment (EIA) process, as it guarantees public involvement and openness in the decision-making process. The public hearing allows the project proponent to respond to stakeholders' concerns and questions, as well as to assess the potential environmental and community implications of the proposed project.

6.2.1 *Need of Public Hearing*

Public hearings are a crucial component of the Environmental Impact Assessment (EIA) process. A public hearing is a platform where individuals from the public can express their feedback and viewpoints regarding a proposed project and its possible environmental, social, and economic consequences. Public hearings in the EIA process are crucial for transparency, accountability, and engaging stakeholders in decision-making.

Transparency is a key factor in the importance of public hearings within the EIA process. Public hearings provide the public with information regarding the proposed project and its potential environmental and social consequences. The public can make informed decisions regarding the project and its potential consequences on their lives and communities by receiving this information. Public hearings enhance the transparency and accountability of the Environmental Impact Assessment (EIA) process by providing the public with the opportunity to express their ideas and concerns on the project. Public hearings are essential for engaging stakeholders in the decision-making process. Stakeholders are individuals or groups who could be impacted by the project, such as local communities, non-governmental organizations (NGOs), and government agencies. Public hearings allow stakeholders to voice their perspectives on the proposed project, perhaps impacting the decision-making process.

Public hearings serve as a method of conflict resolution, along with promoting transparency and involving stakeholders. Public hearings offer a platform for stakeholders to express their issues and come to an agreement on how to deal with them. The EIA process frequently entails conflicting interests, and public hearings can assist in identifying and resolving these conflicts by convening stakeholders to express their concerns and collaborate on finding a resolution.

Multiple studies have emphasized the significance of public involvement in the Environmental Impact Assessment (EIA) process and the necessity of public hearings. A research by Kameswara Rao et al. (2014) shown that public hearings were essential for public participation in the EIA process in India, ensuring transparency and accountability. Araral et al. (2014) conducted a study on public engagement in the EIA process in the Philippines. They found that public hearings were crucial for enhancing transparency, accountability, and responsiveness to stakeholders' demands and concerns.

6.2.2 Role of Local Residents in Public Hearing

During the EIA process, public hearings are held to let local citizens and stakeholders to voice their opinions and concerns about the proposed project. Local citizens' participation in public hearings is essential since they are the most impacted by the project and possess the most knowledge about the local environment and community. Local communities can play a crucial role in detecting potential impacts of the planned project that may have been missed in the EIA report. They can offer important insights about the local environment, traditional land use practices, and the potential effects of the project on their way of life.

Local individuals can utilize the public hearing to express their worries and thoughts on the project, aiding decision-makers in comprehending the social, cultural, and economic environment in which the project will take place. Involvement of local inhabitants can foster trust between the project proponent and the local community, crucial for the project's effective execution.

Furthermore, the involvement of community members in public hearings helps guarantee transparency and active engagement in the Environmental Impact Assessment (EIA) process. This can enhance confidence in the decision-making process and improve the chances of addressing the concerns of all stakeholders. Local citizens' participation in public hearings is essential to ensure a comprehensive understanding of the potential implications of the proposed project and to consider the concerns and opinions of the local community in the decision-making process.

6.2.3 Role of Local Associations in Public Hearing

Local associations, including community-based organizations, non-governmental organizations, and civil society groups, can have a significant impact on public hearings during the Environmental Impact Assessment (EIA) process. These organizations embody the shared interests and worries of the local community and can act as a spokesperson for individuals who are unable to engage directly in the hearing.

The role of local associations in public hearings can vary depending on the context and the specific issues at hand. Some of the key roles that local associations can play include:

Advocacy: Local associations can advocate for the interests of the local community and provide a platform for community members to voice their concerns and opinions. They can also work to ensure that the EIA process is transparent and participatory.

Information sharing: Local associations can provide information on the potential impacts of the proposed project and on the EIA process itself. They can also help to translate technical information into a language that is understandable and accessible to the local community.

Mobilization: Local associations can mobilize the local community to participate in the public hearing and to raise awareness about the potential impacts of the proposed project. They can also work to build alliances with other groups and organizations that share similar concerns.

Expertise: Local associations can bring expertise and knowledge on specific issues related to the project, such as traditional land use practices, cultural heritage, or environmental conservation.

Monitoring: Local associations can monitor the implementation of the project and ensure that the conditions of the EIA approval are being met. They can also provide feedback on the effectiveness of the mitigation measures and recommend improvements if necessary.

Overall, local associations can play a critical role in ensuring that the voices of the local community are heard in the EIA process and that the potential impacts of the proposed project are fully understood and addressed.

6.2.4 Role of NGOs and Other Environmental Groups

NGOs and environmental groups can have a crucial impact on public hearings in the Environmental Impact Assessment (EIA) process. These groups can advocate for public interests, provide specialized knowledge, and enhance transparency and accountability in the decision-making process.

The role of NGOs and environmental groups in public hearings can include:

Advocacy: NGOs and environmental groups can advocate for the interests of the local community and the environment. They can raise concerns about the potential impacts of the proposed project, provide recommendations for mitigating these impacts, and ensure that decision-makers hear the public's concerns.

Expertise: NGOs and environmental groups can provide technical expertise and knowledge on issues related to the proposed project. This can include information on environmental impacts, social and economic impacts, and cultural heritage.

Information sharing: NGOs and environmental groups can provide information to the public about the proposed project, the EIA process, and the potential impacts. They can also help to translate technical information into a language that is understandable and accessible to the public.

Monitoring: NGOs and environmental groups can monitor the implementation of the proposed project and ensure that any mitigation measures are being properly implemented. They can also provide feedback on the effectiveness of these measures and recommend improvements if necessary.

Legal support: NGOs and environmental groups can provide legal support to the public and help ensure that the decision-making process is transparent and accountable. This can include challenging decisions that are not consistent with the law or with the findings of the EIA.

NGOs and other environmental groups are crucial in informing and engaging the public, ensuring they have a voice in the decision-making process. They can assist in ensuring that the potential consequences of the planned project are comprehensively analyzed and resolved, while safeguarding the welfare of the public and the environment.

Summary

The Environmental Impact Statement is an essential element of the Environmental Impact Assessment process. It equips decision-makers with the essential knowledge to evaluate the viability, environmental impact, and societal approval of the project under consideration. An extensive Environmental Impact Statement (EIS), following national and international norms, is crucial for properly addressing environmental and socioeconomic factors. Public hearings in the EIA process are essential. Public hearings are crucial for promoting transparency, engaging stakeholders, resolving conflicts, and guaranteeing accountability in the decision-making process. Therefore, it is crucial that public hearings are carried out in a proficient and productive manner, ensuring that stakeholders have the required information and resources to engage fully in the decision-making process.

Questions/Exercise

1. What do you understand from Environmental Impact statement?
2. How EIS plays an important role in overall EIA process?
3. What is public hearing in EIA process?

References

Araral E, Shen L, Chua L (2014) Public participation in environmental impact assessment: insights from the Philippines. J Environ Manag 135:10–19

CEQ (Council on Environmental Quality) (2023) The National Environmental Policy Act. https://www.govinfo.gov/content/pkg/COMPS-10352/pdf/COMPS-10352.pdf.

EPA (U.S. Environmental Protection Agency) (2023) Environmental Impact Statement. https://www.epa.gov/nepa.

Glasson J, Therivel R, Chadwick A (2013) Introduction to environmental impact assessment, 4th edn. Routledge, London

Kameswara Rao K, Jain AK, Jha A (2014) Public participation in environmental impact assessment in India: current scenario and future challenges. J Environ Manag 143:154–164

Chapter 7
EIA Procedure—Decision Making

The Environmental Impact Assessment (EIA) process was established to ensure that environmental factors are taken into account when making decisions on significant proposals. Those engaged in EIA need to comprehend how the decision-making process functions and the specific role that EIA plays. EIA serves as a process for gathering information to support environmentally conscious decision-making. This process leads to a definitive verdict on the proposal's acceptability and outlines any accompanying conditions. While 'decision-making' in EIA generally refers to the ultimate approval of a proposal, several 'interim' decisions are made during the EIA process. These decisions include selecting a preferred alternative and implementing planning and design adjustments to the initial proposal. In various EIA setups, different entities may be responsible for making interim decisions. For instance, at the screening and scoping phases, the responsible authority typically determines the fate of the proposal. In the EIA preparation stage, the proponent often adjusts the proposal to enhance its environmental and social suitability. The ultimate approval of significant proposals is commonly a political decision, made by the national government, planning authority, or a similar body. In certain EIA systems, approval is necessary before proceeding with other steps. The detailed flow chart showing the process of decision making is depicted in Fig. 7.1.

7.1 Importance of Decision Making in EIA

Sustainability practices are increasingly recognized around the world due to their ability to both support economic development and protect the environment (Esseghir and Khouni 2014). As part of the "US National Environmental Policy Act" (US NEPA) and the "EU Environmental Impact Assessment Directive" (Bond et al. 2020), ex-ante environmental evaluation procedures are important mechanisms to ensure that economic growth is environmentally sustainable. In 2014, European Union (EU)

© The Author(s), under exclusive license to Springer Nature Switzerland AG 2024

R. Bhateria et al., *Environmental Impact Assessment*,
https://doi.org/10.1007/978-3-031-66797-8_7

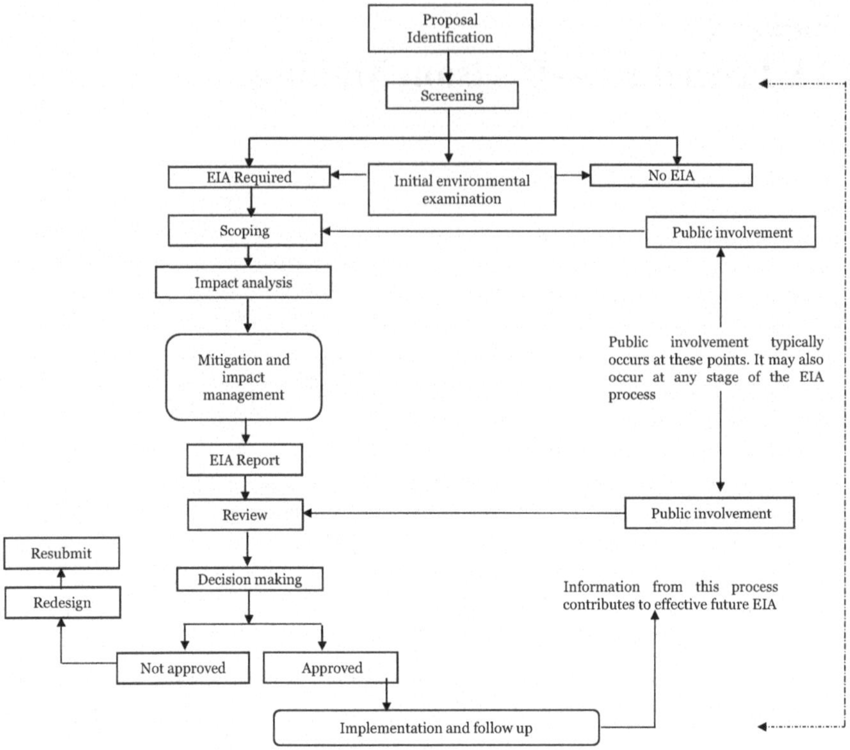

Fig. 7.1 Decision making process in EIA

introduced the Environmental Impact Assessment (EIA) procedure which acts an important tool in preventing negative impacts of project on environment. EIA practices are shaped by relevant models like US National Environmental Policy Act and EU EIA Directive in both national and transboundary contexts (Morgan 2012). A country's core EIA procedures as well as its implementation quality may differ from project to project, when compared to other countries. Some issues, however, obstruct the proper implementation of environmental impact assessments, including regulations specific to EIAs at the national and regional levels, lack of public participation, inadequate trained staff and equipment, lack of cooperation among researchers, policymakers and stakeholders and insufficient institutional framework.

Several emerging economies including China, India, and Brazil developed EIA procedures and governance systems around the world as a result of their own unique sustainable development challenges. Many internal and external factors must be considered by the decision maker during the decision-making process of EIA, according to studies conducted by academicians. They studied technical decision-making as well as business decisions, for instance whenever a decision is made three criteria (technical, managerial, and social) must be taken into consideration. It is important to consider legal restrictions, standards, codes, and previous precedents when selecting technical criteria. Human values, cultural issues, ethical concerns, and environmental impacts are social criteria. Several management criteria including equipment, costs, training, personnel, and demand are also considered while decision making. However, to make an environmentally sound decision, it is ideal for the agency decision-maker to consider all three criteria mentioned above. The process of making environmental decisions is also considered complex by academics. Various interests and values are at play in environmental decisions which make them scientific and technical as well as social, cultural, and economic. According to Stern and Dietz (2008), sometimes environmental decisions present certain complex choices to choose among values and interests. A defensible and informed decision is the eventual objective of an environmental assessment. The achievement of this goal becomes difficult due to certain hindrances such as multiple stakeholder pressures and objectives, the conflicting constraints between environmental options, and the amount of information that must be considered by the public and decision-makers regarding the project. Consequently, "environmental assessment decisions" represent the broad category of decisions involving multiple criteria and objectives. Information presented in an EIS may not be used by agency decision makers. The conclusions of an EIA do not affect the decisions made by the decision makers in particular projects or specific agencies, according to various scholars. Processes influence decision makers instead of decision makers influencing processes.

7.2 Role of Environment Appraisal Committee in Decision Making

A development/infrastructure project is cleared by the Ministry of Environment based on recommendations from expert committees appointed by the Ministry (currently eight such committees exist) and Expert Appraisal Committee (EACs). Let us see an example where the decision raises many questions, particularly from the perspective of civil society involvement during the environmental clearance process. When the Environment Ministry's EAC for Hydroelectric Projects and River Valley met on 30 December 2016, it decided that civil society groups would not be given consideration when assessing the final appraisal of an environmental clearance application for the proposed project. According to the meeting minutes, many of these representations lack a development perspective, thereby causing delays or suspensions in the projects. Developers in particular and the nation in general will be affected financially by the decision. Following the completion of the public consultation stage, the EAC considered receiving representations that would mean reopening the process once again and it will be judged to be satisfactorily completed.

The project proponent prepares a draft EIA report based on the Terms of Reference (ToR) and other relevant documents which is then posted in public domain for public consultation (local and regional governments' offices and official websites). Projects are listed in the EIA Notification that are exempted from public consultation. Proponents are anticipated to respond to concerns raised during the public consultation and finalize the draft EIA report on the basis of results of the consultation. As part of its evaluation process, EACs review the EIA report, public consultation outcomes and other relevant documents. In recommending Environmental Clearance (EC), the EAC needs to carefully review the documents and the presentation of the project proponent, and then decide whether to recommend EC (or rejection) or additional studies are required to take a decision. Final decisions are made by the Ministry of Environment, Forest and Climate Change (MoEFCC) based on recommendations from the EAC. It is usually accepted by the MoEFCC, but it can request the EAC to re-evaluate its recommendations if it disagrees. However possible impacts of project are considered and the EAC review project proposals and make recommendations to the MoEFCC and it decides either rejects or grants clearance with certain conditions that mitigate or compensate for the impacts based on these recommendations.

EC team plays a crucial role in the EC process, and it is important that they are independent and its objective for them to bring expertise. An EC process has several flaws, the most critical of which is the fact that the project proponent pays for the EIA. It is essential to verify the accuracy, assumptions, and safeguards of the proposed project by the proponent since significant information about the project and its probable impacts is provided. Although civil society groups are important stakeholders in the EC process, the minutes of the EAC for River Valley and Hydroelectric Projects seem biased against them. In their objections, alternative assessments, and criticisms of the project proposals, they may highlight errors or omissions. As long as all stakeholders' opinions are taken into consideration, the EACs should keep an open mind.

In addition, the MoEFCC has a duty to notify the EACs of all public comments received on proposed projects by the MoEFCC, according to the EIA Notification 2006. These representations are neither to be assessed nor to be determined to be appropriate. A public representation cannot be taken into consideration by the EAC if it is repetitive, "does not raise new issues, or is unsubstantiated" unless they have discretionary authority to do so. A final consideration is the EAC members' expertise and experience in environmental assessment and protection. A proposal for a project and its impact, as well as mitigation measures are addressed by an expert opinion provided to the MoEFCC. EACs should not consider financial implications or delays in project approval when deciding whether to approve a project. By sympathizing with one stakeholder (the developer) while rejecting both the public's and the developer's concerns, they compromise their neutrality.

An EC must be obtained by certain project categories prior to proceeding with project work based on a Notification issued by the Central Government under the Environment (Protection) Act 1986. Since EIA studies are the fundamental section of EC process, this Notification is commonly known as the EIA Notification 2006. The list in Schedule to the EIA Notification consist types of projects that require prior ECs which includes river valley power/irrigation, thermal power projects, mining, airports, industries, highways and solid waste management. Each state's Central Government establishes a State Environmental Impact Assessment Authority

(SEIAA) for this purpose, depending on the size and scope of these projects. The Environmental Clearance is obtained either from the MoEFCC or the SEIAA.

An EC application along with appropriate information about the proposed project is presented by the project proponent to the related regulatory authority, it can be MoEFCC or the SEIAA. In the next step, the project proponent is required to issue detailed and comprehensive Terms of References for preparing an EIA report. EACs can amend standardized terms of Reference for particular projects when specific issues arise. It is essential for public participation to enhance the legitimacy, transparency, and accountability of "environmental decision-making" processes and, as a result, to make the outcomes more acceptable.

In the EC process, public participation is primarily focused on consultative processes. It is a two-step public consultation process: (1) A public hearing is conducted at the project site or a nearby location to address concerns raised by individuals invested in the project's environmental aspects and (2) Responses in writing from individuals invested in those environmental factors. Being a part of engagement process, the public can also write to the relevant EAC and express specific concerns. Public representations could capture the EAC's focus on particular concerns that were not adequately addressed before, especially if new information has surfaced or there is limited time for thorough consideration. As of the time of the public consultation, there is only the draft of the EIA report available. In response to material concerns raised during the public consultation, the proponent is required to complete the EIA report. However, the public is unable to review the final EIA report to verify if their concerns have been addressed. In assessing whether the final project proposal adequately addresses the public's concerns, the EAC is expected to be the neutral arbiter. The EAC must consider adequate mitigation and ameliorative measures in its recommendations to assuage the concerns of the public if the project is recommended for clearance.

Considering the environmental and social impacts of a project is a key responsibility of the Environment and Advisory Committee. However, financial implications may also arise from this consideration. The terms of reference (ToRs) for members do not specify whether the EACs can take into account public representations sent directly to them or after public consultations have ended. Despite this, the EACs have accepted such representations in the past. EACs overseeing thermal and coal mining projects have expressed concerns about receiving public comments late, which limited their time to address important issues. They requested guidelines from the MoEFCC that would require stakeholders to submit objections a week before EAC meetings, respecting everyone's right to express concerns. In a subsequent EAC meeting (Box 1), it was decided that project proponents must respond to public representations before the next meeting, to allow EAC members to review both the responses and the MoEFCC's comments. It is evident that public representations play a significant role in the EAC's decision-making process.

The Terms of Reference (ToR) for the EAC could be updated to include instructions to take into account public feedback, especially if the MoEFCC is the appointing authority. However, it seems unlikely that these changes will be made, as the environment minister has purportedly directed the experts to expedite project approvals without hindering business operations. According to the Ministry of Environment and Forests (MoEF), industrial projects should undergo environmental assessments promptly.

Box: 1 Expert Appraisal Committee

- EAC meetings should be held at least once every 15 days, according to the recommendations.
- Ten days before a meeting, all proposals that had been submitted for approval should be considered (it has now been 15 days).
- A total of seven EACs are currently operating under the MoEFCC.
- To provide guidance to the government regarding environmental clearance for new development projects, expert appraisal committees (EAC) exist at both the state and the federal levels.
- Obtaining environmental clearance for development projects requires the involvement of the Environmental Assessment Committee.
- An Expert Appraisal Committee (EAC) reviews and recommends methods to mitigate potential environmental impacts of projects beyond a certain size.
- MoEF relies on their advice in deciding whether to approve or red flag projects.
- Screening, scoping, public hearings, and appraisal are the four steps involved in the evaluation of proposals under the Environment Impact Assessment (EIA) Notification, 2006.
- All stages except public hearings are covered by the EAC/SEAC.

November 2020, a new memo was issued by the Union Environment Ministry, stating expert appraisal committee will meet twice a month to "shorten the processing time" for environmental clearances. The EAC is a ministry-wide expert panel that provides approvals or rejections for proposed projects as part of the environment impact assessment process by awarding or withholding environmental permits. Due to its expertise, the Union Environment Ministry in most cases follows the recommendations of the EAC, even though it is an advisory body. Considering this, there is concern that more meetings will result in more approvals of ecologically destructive projects. This commission has a monumental task: determining the terms of reference under which environmental impact assessments (EIA) are prepared, reviewed, and decided whether to approve a project based on these EIA reports and public hearing results or not. The committee's lack of time to critically assess the environmental impact of different projects is also due to this volume of decisions. Secondly, if we want to get a truer understanding of this issue, we need to ask whether members of the EAC are really 'experts' who can make serious environmental decisions independently and objectively. Since the word 'expert' occupies a significant position in the EAC, it is important to recognize this.

7.3 What is NOC?

According to the Water Act 1974 and Air Act 1981, the National Oceanic and Atmospheric Administration requires obtaining Consent to Establish from the respective State Pollution Control Board before construction activities begin. An application in a prescribed format, along with required documents and scrutiny fees, must be filed with the State Pollution Control Board for consent to establish. Afterwards, the site is physically inspected, and the proposed environmental management system is assessed to determine whether it meets the mandatory requirements imposed by the State Pollution Control Board. States differ in how much data they require to carry out the Consent to Establish. As the time to obtain the NOC from the concerned authority varies according to the type and area of the industry or process, the time for obtaining the NOC is also influenced by the type of industry, since green businesses get the NOC faster than orange or red businesses. In addition to the project location, the length of time it takes to obtain consent to establish depends on the condition of the project. After the commencement of development activities, the respective State pollution Control Board must be contacted to receive Consent to Operate (CTO). A similar set of documents is required for the process of obtaining the Consent to Operate.

7.3.1 Steps to Be Followed for the EIA NOC

- Preparation of proposal: Assist in the preparation of the proposal by compiling information from the proponent and the architect to prepare a detailed project report.
- Expert assessment of projects: Expert assessment of project reports by environmental experts in different functional areas
- In the process of assessing ambient environmental conditions, accessibility to the proposed site, and environmental impacts, a site visit is an important step.
- As per CPCB (Central Pollution Control Board) and SPCB (State Pollution Control Board) and MOEF norms, the NABL accredited lab will collect baseline data on Air, Water, Noise, and soil of the site before and after the monsoon, and any other parameter that is applicable.
- A final proposal submission is completed and submitted to SEAC (State Expert Appraisal Committee), SEIAA (State Environmental Impact Assessment Authority), and other relevant agencies.
- Presentation to SEAC and SEIAA: Presentation will be made to SEAC and SEIAA
- An EC NOC is issued after the EIA letter has been awarded, and after the project's impacts, mitigation measures, and other technical and legal parameters have been assessed.

7.3.2 *Validity of the EIA NOC*

- The EC NOC remains valid for seven years for projects related to the development of buildings and townships. Before the expiration of the EIA NOC, the DPR (detailed project report) can be submitted to the Environment Department if the EC NOC letter validity has expired.
- Compliance reports should be submitted six months after the EIA
- There is a mandatory requirement to submit a post environmental clearance report to the Environment Ministry and other relevant authorities twice each in June and December every year, irrespective of the schedule notified under the EIA notification 2006 and subsequent amendments by the Ministry of Environment and Forest.

7.3.3 *Requirements for Obtaining EIA NOC*

- An impact assessment report containing details about the proposed activities which would have significant and non-significant impacts on fauna and flora, as well as environmental aspects such as air quality, water quality, geology, and hydrology, is submitted to the Environment Ministry when a project is obtaining an Environment NOC from the Ministry of Environment. The mitigation measures taken to address them.
- The Project proponent agrees to comply with all the conditions outlined in the EIA NOC based on the Environment Impact Assessment report and conceptual report submitted. The environment ministry won't know how the project proponent complies with the impact mitigation measures committed to in the detailed project report when that letter is given, until the project's impact mitigation measures are actually implemented during construction and thereafter during operation.
- To comply with environmental regulations, a six-monthly compliance report is required to demonstrate the project's growth while remaining in compliance with recent environmental techniques and pollution control measures.
- The projects and activities under the EIA notification 2006 must follow the conditions of the EIA NOC, and the EC letter must be accompanied by a detailed project report.
- In every case, if the project has been granted EIA NOC, it has to be advertised in newspapers in Hindi and English and in local and regional papers.

7.4 Environmental Clearance (EC) Process

The EC process is mandatory for 39 project types, involving screening, scoping, and evaluating the upcoming project. Its primary goal is to assess how the project affects the environment and individuals, aiming to reduce or mitigate any negative effects. The projects have been characterized into 2 major categories summarized as:

Category A: It is managed at the central level. Every project must have an Environmental Impact Assessment (EIA) report for evaluation and go through a public consultation process.

Category B: The category 'B' is subdivided into 'B1' and 'B2', excluding township and area development projects in category 8a of the schedule.

Category B1: The projects classified as B1 at the State level necessitate an EIA report for evaluation and must also go through a public consultation process. All projects/activities under Category 'B' as outlined in the Schedule of EIA Notification, 2006, and its revisions, are to be regarded as Category 'B1' projects and assessed according to the procedures detailed in the EIA Notification, unless explicitly specified in Category 'B2'.

Category B2: The projects classified as 'B2' are evaluated at the state level or district level, depending on the sub-category. All projects, except for minor mineral mining projects, require an appraisal based on the application, pre-feasibility report, and other necessary documents at the state level.

However, to obtain Environmental Clearance, the following steps must be taken:

- The project proponent selects the planned plant site once ensuring it complies with current siting guidelines. If the site does not adhere to the guidelines, the proponent must locate a different site.
- The project proponent first assesses if the proposed activity/project needs environmental clearance by checking the notification schedule. If listed, an EIA study is conducted by the proponent alone or with a consultant. Projects under category B require state government clearance, which is divided into B1 and B2 projects. EIA report preparation is not necessary for B2 projects.
- Once the EIA report is finalized, the stakeholder engages with the relevant State Pollution Control Board (SPCB) and, if applicable, the State Forest Department (in cases involving forestland use). The SPCB reviews the projected effluent quantity and quality from the proposed unit, along with the adequacy of investor's proposed control measures to meet the specified standards. Upon satisfaction with the compliance to effluent and emissions standards, the SPCB grants the "Consent to Establish" (also called as "NOC") which remains valid for 15 years.
- Public hearings play a crucial role in the environmental clearance process for certain development projects, enabling local residents to express their concerns directly to both the project proponent and government officials.
- The public hearing occurs prior to obtaining a No Objection Certificate (NOC) from the State Pollution Control Board (SPCB). Chaired by the District Collector,

the public hearing committee includes members from the SPCB, district development body, representatives from Gram and Taluka Panchayat, Department of Environment and Forest, and senior citizens of the district. During the session, the committee reviews public objections and suggestions, incorporates specific provisions, and then submits it to the Ministry of Forest and Environment for approval.

- Projects falling under Category A require the project proponent to submit an environmental clearance application to the MoEF. However, B category projects should send their application to the state government. The application needs to contain the EMP, NOC from state regulators, information about the public hearing, and the EIA report.
- Investors' documents undergo thorough examination by a diverse team within the MoEF. This team may visit sites, engage with investors, and seek advice from experts on particular matters. After initial review, the proposals are presented to specialized expert committees as outlined in the EIA Notification. Environmental Appraisal committees are formulated for each sector and for controversial projects or special projects, public hearings may also be arranged to make sure of public participation. Public hearings must be announced at least 30 days in advance through newspapers. Following the outlined process, the Appraisal Committees provide recommendations for approving or rejecting specific projects. The Ministry of Environment and Forests then reviews and decides on the Committees' recommendations for final approval or rejection.
- In cases involving clearance or rejection letters for projects needing both environmental clearance and approval under the "Forest (Conservation) Act, 1980", these proposals must be submitted at once to the relevant departments of the ministry. The processing is carried out concurrently for clearance or rejection, although separate letters may be issued. If the project does not entail the diversion of forest land, only environmental clearance processing is undertaken.

Upon receiving all necessary documents and data from the project authorities and conducting any required public hearings, the environmental evaluation of the project is finalized within 90 days. The decision made by the ministry will be communicated within a timeline of 30 days thereafter. The validity of granted clearance lasts upto 5 years for initiating project construction or operation. A step-by-step schematic flow chart showing environmental clearance process in India is represented in Fig. 7.2.

7.5 Implementation and Follow up in EIA Process

A critical yet relatively overlooked step of the EIA process is implementing and following up on the EIA. A proposal can be assessed and reviewed in an ongoing manner by monitoring, auditing, evaluating, and other methods following its final approval through surveillance, monitoring, and auditing. As part of the EIA process, impacts are identified, checked whether they are within the range predicted and

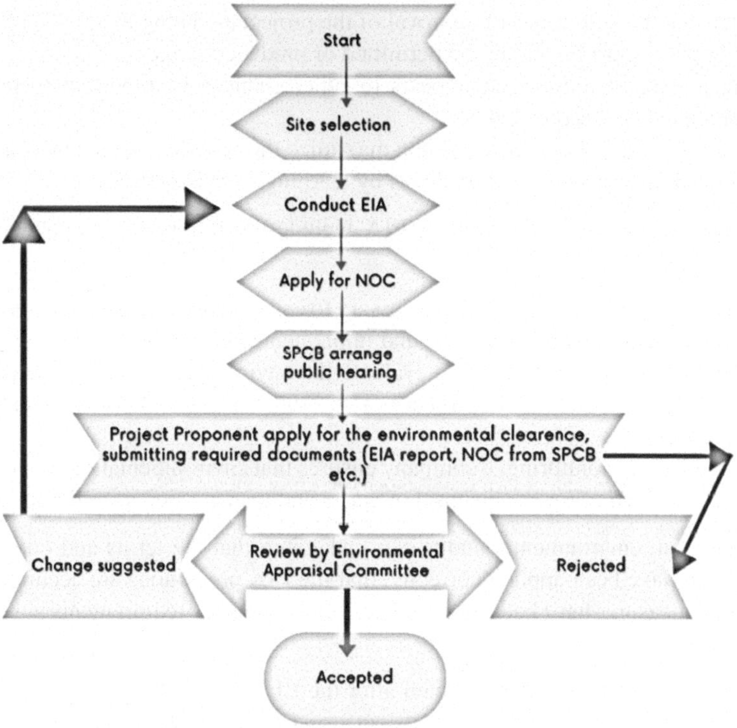

Fig. 7.2 Environmental clearance process in India

required by law, mitigation measures are determined, and feedback is provided to improve future applications of the process. In addition to ensuring the environmental benefits expected are being achieved, they provide feedback to help improve future applications. Project construction and operations have received relatively little attention until recently. A decision cannot be implemented or followed up if it is not appropriately implemented. Instead of achieving environmental benefits, EIAs become paper exercises for securing approval. Implementing and following up on EIA is important for ensuring that the conditions attached to project approval are fulfilled and working effectively, and to learn from the process so that Future EIA practice can be improved.

Environmentally unsound projects cannot be turned around by this process alone. EIA reports should, however, be prepared and considered in decision making in order to maximize their benefits. By implementing and following up on EIAs, new information can be used to refine project approval measures and conditions. As part of an EIA process, they facilitate a systematic approach to impact management that ensures continuity and optimizes environmental benefits throughout the project lifecycle. Implementing and following up on EIAs must achieve the following objectives:

- Verify that the conditions of approval of the project are being met.
- The impacts must be within the permitted or predicted ranges.
- Management of unforeseen impacts or other changes is important to prevent unanticipated consequences.
- By following good practice, you can maximize the benefits for the environment.
- Improve EIA processes and practices by learning from experience.

Implementation and follow-up of EIA include the following components and tools:

- Supervision and surveillance are necessary to ensure that the terms and conditions of project approval are adhered to and implemented correctly.
- Effect or impact monitoring—to evaluate the environment's changes caused by the building or operation of a project and to ensure mitigation measures effectively reduce them.
- In compliance monitoring, a company ensures that environmental regulations are being met, such as for the disposal of waste and the emission of pollutants.

In addition, environmental audits can make sure that the terms and conditions of the grant have been applied correctly, that the EIA predictions are accurate, that mitigation measures have been effective, and that regulatory requirements have been met.

- An ex-post evaluation is conducted after the EIA process has been applied to a specific project in order to review its performance and effectiveness.
- An overview of the project development and lessons learned will be presented in the post-project analysis.

Each country establishes institutional arrangements and procedures for this purpose that define and delineate these components in different ways. They are, however, relatively well understood in terms of their generic functions. As part of EIA implementation and follow-up, several monitoring, auditing, and evaluation techniques are described in the accompanying box. During the design, construction, and operational phases of the project, these environmental parameters are analyzed in accordance with EIA, EM&A procedures, as shown in Table 7.1.

In judging whether a proposal will be approved, the following factors will be considered:

- According to the EIA report, there were significant impacts.
- A summary of interim stages' inputs;
- Decisions are influenced by other factors, such as external pressures or political influences

Table 7.1 Showing stages of decision making and related information

Decision-making stage	Important information
Background	Project background and the most important environmental issues involved
Policy context	Basic development issue or problem being addressed (e.g., flooding, water shortage, etc.); The relationship to environmental policies and plans
Alternatives	Alternatives to the proposal (including the best practical environmental option (BPEO) or equivalent designation)
Public involvement	Key public views; Concerns of affected communities; Areas of agreement and disagreement
Impact analysis	Distribution of gains and losses
	Costs and benefits
Mitigation and monitoring	Adequacy of proposed measures
Conclusion and recommendations	Main economic benefits, significant environmental effects, and proposed mitigation measures
	The extent to which the proposal conforms to the principles of sustainable development
	Design and operational changes to improve the environmental acceptability of the project

Summary

Environmental Impact Assessment (EIA) is commonly viewed as a tool to assist authorities in making informed decisions regarding project approval and necessary conditions. Decision-making occurs throughout the EIA process where proponent typically makes several decisions. Certain decisions are collaborative efforts between the proponent and environmental authorities, like screening and scoping decisions. The primary decision in the EIA process, whether to permit the proposal to proceed, rests with a government agency after consultation and public involvement. At this stage, the decision often involves choosing between authorization, conditional authorization, or refusal, rather than selecting between alternatives. This section examines the decision-making procedures employed by government agencies when approving or rejecting projects with significant environmental impacts. While one might expect agencies to follow a well-defined process, external and internal factors play a substantial role in influencing the decision-making process.

Questions/Exercise

1. How are the EIA process and EIA document used in the decision-making process.
2. Describe the role of environmental appraisal committee in decision making.
3. Enlist points to consider the importance of public consultation in decision making process of EIA.
4. Outline the factors important for decision making process.

References

Bond A, Pope J, Fundingsland M, Morrison-Saunders A, Retief F, Hauptfleisch M (2020) Explaining the political nature of environmental impact assessment (EIA): a neo-Gramscian perspective. J Clean Prod 244:118694

Esseghir A, Khouni LH (2014) Economic growth, energy consumption and sustainable development: The case of the Union for the Mediterranean countries. Energy 71:218–225

https://ebooks.inflibnet.ac.in/esp12/chapter/environment-clearance-process-in-india/

https://www.cseindia.org/environmental-clearance---the-process-403

https://www.fao.org/3/V8350E/v8350e06.html

https://www.iaia.org/pdf/south-eastern-europe/Sec_E_Topic_10.PDF

Morgan RK (2012) Environmental impact assessment: the state of the art. Impact Assess Proj Apprais 30(1):5–14

Stern PC, Dietz T (2008) Public participation in environmental assessment and decision making. National Academies Press, Washington, DC

Chapter 8
EIA Procedure—Post Monitoring and Environmental Audit

Once a project is implemented, monitoring plays a vital role. It ensures that the project's impacts adhere to legal standards and that mitigation measures are implemented as outlined in the Environmental Impact Assessment (EIA) report. Monitoring also gathers data on the environmental and social effects of the project throughout its lifecycle. The data collected during monitoring activities helps confirm the proper implementation of priorities listed in the Environmental Management Plan (EMP), mitigation measures, and contingency plans. It is crucial for upholding commitments outlined in the approved EIA and for tracking environmental and community changes due to the project or other local/global events. Basic monitoring continues post-project implementation and during project remediation. Additionally, an environmental audit evaluates a business or organization's environmental performance, compliance with environmental regulations, and management system implementation. The audit provides insight into a company's activities and environmental compliance to management and staff. It also identifies compliance issues and management system implementation problems. The audit helps in measuring the environmental impact of specific issues or activities such as air quality, materials management, or wastewater. A functional environmental audit focuses on particular concerns and may be part of an environmental compliance or management audit.

8.1 Monitoring in EIA

A monitoring program compares the conditions at a project site before and after the project has been completed. Additionally, actual impacts are compared with those predicted. This is especially true when it comes to key impacts like water quality, air quality, soil fertility, endangered species, etc. Moreover, monitoring helps ensure that government regulations and laws regarding a development project are being followed by authorities involved in it. In this way, all development authorities are

© The Author(s), under exclusive license to Springer Nature Switzerland AG 2024

R. Bhateria et al., *Environmental Impact Assessment*,
https://doi.org/10.1007/978-3-031-66797-8_8

discouraged from cutting corners. In addition, monitoring has the bonus of identifying potential impacts in case the EAC missed them at some point. As a matter of fact, this is quite possible. It is quite possible that the team missed some of the impacts that could have been caused by the activity because of the multitude of effects that a single activity can have. Observing significant environmental damage permits project proponents to alter their activities. It is important to understand that prediction studies are only theoretical, so there is a high probability of errors. Making appropriate corrections becomes possible when results are observed in real-time. Monitoring can be of following types:

***Impact/effect monitoring*—Monitoring the impact and effect of environmental impacts—analyzing these two effects in direct relation to each other.** Intensity, magnitude, and extent are included in this type of monitoring.

***Baseline Monitoring*—In baseline monitoring, an evaluation is conducted before a project begins.** Data from baseline monitoring is used to determine the site's readiness for project development. Before finalizing the commissioning of the project, the team needs to gather baseline data on air quality at all key locations on the site for 14 consecutive days. This process allows them to obtain dust samples every 24 h.

***Mitigation monitoring*—This involves determining whether the measures are effective.** All mitigation measures are assessed, whether they are implemented, designed, or technological in nature.

***Compliance Monitoring*—**This approach involves thoroughly assessing the alignment between predicted and actual values, encompassing mitigation monitoring, impact assessment, and effect monitoring.

All stakeholders involved in the EIA process can benefit from monitoring and management response, which can improve several aspects of EIA practices. Environmental impacts of development projects are becoming increasingly recognized as affecting corporate reputation. Developers can demonstrate environmental responsibility and confirm that their development protects the environment by monitoring their progress post-decision in EIA.

8.2 Environmental Audit

Environmental auditing traces back to the United States in the 1970s. The pollution attributed to oil and chemical companies during that era sparked the necessity for environmental audits. In 1979, the US Environmental Protection Agency (EPA) introduced a preliminary report, inviting independent, certified third-party auditors to inspect plants, gather samples, conduct analyses, and present reports to government bodies. This governmental auditing approach garnered notable interest but faced strong resistance from the industry. Environmental auditing in Europe originated from the significant environmental risks posed by chemical and petrochemical industries on a large scale. Since the 1980s, the idea of environmental auditing has gained widespread acceptance among industries as a standard management tool in developed countries. It is now also becoming more prevalent in developing nations. The adoption of environmental auditing stemmed from the dedication to implementing Agenda 21.

Environmental auditing is a systematic, documented, periodic and objective process in assessing an organization's activities and services in relation to:

- Assessing compliance with relevant statutory and internal requirements
- Facilitating management control of environmental practices
- Promoting good environmental management
- Maintaining credibility with the public
- Raising staff awareness and enforcing commitment to departmental environmental policy
- Exploring improvement opportunities
- Establishing the performance baseline for developing an Environmental Management System (EMS)

8.2.1 Steps in Environmental Auditing

The environmental audit process involves the following steps at a minimum:

- Plan the audit, detailing activities and assigning responsibilities
- Review the company's environmental protection policy and relevant regulations at the federal, state, and local levels
- Assess the organization, including management and equipment
- Collect data and pertinent information
- Assess the overall performance
- Identify areas that require improvement
- Present findings to management

8.2.2 Importance of Environmental Audit

Environmental audits evaluate a company's environmental impact, guiding management on enhancing environmental sustainability, identifying processes that boost productivity, and detecting risks proactively. The central objective of environmental audits is to ensure adherence to environmental laws. Each region has its own set of environmental regulations that companies must comply with to avoid penalties. These audits play a critical role in standardizing compliance measures throughout the organization. They help identify areas requiring training, assess the effectiveness of control systems, and evaluate communication channels related to environmental compliance.

8.3 Features of Environmental Audit

- **Systematic—A systematic audit of an environmental condition requires careful planning, structuring, and organization.** Due to its role in evaluating and checking over time, it needs to be a repeatable process so that different teams of people can use it over time in a manner that allows the results to be compared and to reflect qualitative and quantitative changes.
- **Objective Evaluation—An audit is always subject to subjectivity, especially if the audit is conducted internally, or if predetermined policies and procedures are not followed.** Furthermore, reviewing environmental data periodically (say annually) by independent audit teams with specialized skills tends to increase the system's objectivity. Hence, external environmental audits should be used to maintain objectivity other than certification guidelines (such as ISO 14001).
- **Periodic—Audits are generally carried out at predetermined intervals.** The process takes time.
- **Documented—Audit findings should be supported by documents and verifiable information, just as they are in any other audit.** By using available evidence, the audit process verifies past actions, activities, events and procedures on a sample basis to determine whether the procedure was completed in compliance with the system requirements.
- **Environmental Performance—Environmental Audits are intended to determine whether a company or organization's environmental performance is up to par with world standards in terms of environmental management and environmental equipment.** To achieve its goals, a company should align its environmental performance with its policy, compliance standards, and regulatory requirements.

8.4 Objectives of Environmental Audit

- Providing regulatory bodies with a report and achieving compliance standards.
- Evaluation of policy implementation.
- Identifying liabilities. Performing a management system review.
- Identifying strengths, weaknesses, and needs.
- Performance assessment of the environment
- Awareness of environmental issues is promoted.
- Increasing the safety and health of production
- As a means of conserving nocturnal resources.
- As a way to reduce waste,
- A regulatory requirement must be assessed for compliance
- Informing the public about environmental issues.

8.5 Benefits of Environmental Audit

Auditing the environment takes place in the third phase of the cycle, setting the stage for adapting to environmental demands in the fourth phase. Environmental Auditing brings numerous advantages, as outlined by the International Chamber of Commerce, including:

- Interchanging information between operations and plants, facilitating comparisons,
- Awareness-raising among employees concerning environmental policies,
- Assessing cost savings resulting from waste minimization, including those associated with waste reduction,
- Training programmes are evaluated, and data is provided to assist with personnel training.
- Providing emergency response arrangements with an information base,
- Ensuring that an adequate and current environmental database is available to internal management to aid their awareness and decision-making regarding modifications and new plans for the plant, etc.
- Achieving environmental excellence through management credit,
- Establishing a relationship with authorities through convincing them that audits are being conducted in a thorough and effective manner, informing them of the types of audit procedures taken, assisting them in their relations with the authorities,
- Enabling environmental impairment liability insurance to be obtained.

Other than above mentioned benefits, certain other benefits are categorized as:

Financial Benefits

- Reduces the chances of regulatory fines
- A proactive approach to financial planning allows for early detection of issues of non-compliance, resulting in lower corrective action costs
- Reduction of operating costs due to waste minimization opportunities

Compliance Benefits

- Penalties and enforcement actions reduced by the agency
- Environment-related standards and responsibilities raised employee awareness

Stakeholder Relations Benefits

- An improved working environment and a higher level of morale among employees
- The Company's goodwill in the community has been improved
- Investors are attracted to companies that maintain sound environmental stewardship programs

General Benefit

- Maintaining sustainable development levels.

8.6 Types of Environmental Audit

There are two main types of environmental audits:

8.6.1 Environmental Compliance Audit

During an environmental compliance audit, a company's environmental legal requirements are evaluated independently, and assessed for company's compliance. A company can use this tool to measure and evaluate its environmental performance, and to identify problems with compliance or management system implementation. In addition to ensuring compliance with applicable regulations and laws, the environmental compliance audit also ensures that internal guidelines, controls, policies, and procedures are followed by the company. If environmental regulations are not followed, undesirable consequences may occur, including financial penalties and criminal charges. In order to avoid these outcomes, organizations should regularly conduct internal and external compliance audits. During an environmental compliance audit, stakeholders are ensured that the company is operating in an environmentally sustainable manner and that it is transparent with regards to its environmental footprint.

Benefits of Environmental Compliance Audits: There are several reasons why companies should conduct an Environmental Compliance Audit. Some of the benefits are enlisted below:

- Makes sure the environment is protected
- Assesses compliance with national and local laws
- Provides an indication of critical environmental problems and impacts in the near or distant future
- Provides data for environmental training in the future by assessing training programs
- Improves the environmental performance of companies and highlights shortcomings
- Analyzes waste minimization opportunities for savings
- Achieves environmental protection targets for the company and communicates them to employees, other stakeholders, and authorities

Environmental compliance audits are further categorized as:

Environmental Compliance Audit (External): Construction site or company legal compliance audits are the most comprehensive types of environmental audits.

Environmental Management Audit (Internal): Organizations and managers conduct internal audits to verify that their environmental performance goals are being met.

Environmental compliance audits are the most common audits conducted by companies. These audits involve checking specific regulatory requirements using a checklist to ensure the company meets these standards. Each task a company undertakes is scrutinized for compliance during an environmental audit. Depending on the industry and operations, various processes may require auditing. Typically, these audits are necessary for obtaining environmental authorizations and licenses. Evaluating how a company minimizes environmental impact aids in ensuring they are meeting all necessary criteria. The organization providing certifications clearly outlines industry-specific requirements, simplifying compliance checks for companies.

It is usually more comprehensive to conduct environmental compliance audits since they review a company or site's legal compliance status (which is both important and costly). Environmental compliance audits are designed to assess compliance with a specific and defined checklist. As part of a compliance audit, an organization is assessed and verified against a specific objective. Different regulatory approvals include compliance commitments, like the Environmental Authorization or the specific management license (related to waste, water, and the atmosphere). A legal management plan is included in the approval document in order to prevent environmental damage during the activity. As part of this activity, company must comply with these measures, or conditions. Authorization holders must adhere to these conditions as required by the competent authority.

Audits of the environmental impacts of activities are conducted by independent parties with the objective of monitoring and reporting on compliance. A report of the status of complying with all requirements in the authorization is compiled by the auditor upon reviewing all conditions and requirements in the authorization. The result of the audit process including a report detailing conformity with obligation that set for the project and can be submitted to the component authority. This report recommends improvements for the holder of the authorization to help him or her improve overall compliance with requirements. To ensure continuation of the activities and to prevent environmental degradation, compliance audits inherently identify aspects that requires direct intervention whether from a job title perspective or from a management. Whether it involves a regulated process, self-assessment, or a company review of its environmental performance, environmental auditing offers a distinct advantage by enhancing the overall environmental performance of the operation. Through this method, a detailed summary of the audit findings is generated, aiding in the detection of critical issues or areas of worry to prevent, manage, and effectively reduce environmental impacts associated with the operation in the future.

8.6.2 Environmental Performance Audit

Performance auditing is a more recent activity that dates back only to the mid-to-late twentieth century relative to traditional auditing, which is as old as human civilization itself. Modern performance auditing is considered to have begun during the post-Second World War era, the 1950s, the 1960s, and the 1970s. Despite the lack of clarity regarding the timing of performance auditing's origins, Kells and Hodge (2009) note that the United States, Sweden, Japan, Germany, Canada, and the United Kingdom are all viable global origins. Since its origin in the British Commonwealth and Eastern Europe, modern performance auditing has spread to other countries as well. The UK has been considered to be one of the pioneers of performance auditing by Barrett (Former Auditor General of Australia, personal communication). Furthermore, in the 1970s, Australia began performing efficiency audits as a precursor to performance audits. A new public management model (NPM) enables flexibility in inputs and processes, while increasing the focus on monitoring and evaluating performance (outputs and outcomes) in return. Pollitt and Summa (1999) note that performance auditing is closely linked to the new public management model (NPM).

Accountability and learning for improvement are the two main objectives of performance audits. An audit of performance produces accurate and reliable assessments regarding performance with a lack of investigational and reporting limitations that can be used for accountability purposes. After almost a quarter century since an authoritative body proposed the first definition, there have been two additional definitions that provide insight into the nature and identity of performance auditing. The most common performance audit uses an accountability (systemic/administrative) approach and consequently tests the adherence to management criteria known as 'good management'. In a sense, such an approach is no different from financial auditing from a conceptual standpoint. According to Barrett (Former Auditor General of Australia, personal communication), performance audits have evolved considerably since Pollitt made that observation. Analyzing the economic impact of policy objectives. Entry is not always necessary to perform efficiently and effectively.

8.6.3 Functional Environmental Audits

During these audits, companies are assessed for compliance with certain regulations. The waste management plan of a company's jurisdiction can be audited, for example, to ensure compliance. In addition, air quality monitoring audits can be performed as part of these audits. Having an effective and thorough atmospheric emissions management plan ensures that the company's regulations are met. Regulatory boards can also inspect companies' operations before conducting their own inspections, so that companies can ensure they are compliant. In addition to providing companies with specific information, these specialized audits often provide companies with advice for the future. Water or electricity audits are specific audits that focus on one element of the environment or its impact.

To determine whether an organization is in compliance with a specific aspect undertaken by it, functional environmental audits are conducted. For example, an organization may conduct a waste management audit in accordance with a "Waste Management Plan" or conduct an air quality monitoring audit to examine how an Atmospheric Emission Management Plan is being implemented. It plays a crucial role in verification and checks during compliance audits and Environmental Management System audits. The implementation of functional audits can also serve as a means of determining whether a system needs improvement or if specific compliance requirements need to be met. Functional audits are beneficial as they can be merged with other audit programmes or kept as a single assessment focusing on an area that requires managerial attention.

With these audits, changes can be made quickly without waiting for regulatory approvals. Examination of accounting records and assessment of financial statements for the purposes of attesting to financial accountability of accountable entities;

- Assuring that the taxpayers are being treated fairly by the government administration;
- Assessing compliance with relevant laws and regulations by auditing financial systems and transactions;
- Controls and functions of internal audit are audited;
- A review of the legality and propriety of all administrative decisions made within the audited organization,
- Reporting of any additional information, the Supreme Audit Institution (SAI) believes should be disclosed arising from the audit.

A financial statement offers details about a company's financial status, achievements, and cash flow, aiding in resource distribution. Public sector financial statements must provide decision-makers with useful information and demonstrate a company's financial accountability. Governments (or their constituent entities) typically avoid issues related to the environment in their financial statements. However, financial statements should take into account environmental policies and obligations with regards to costs, compliance, and performance. Information about finances is traditionally the focus of financial auditing. It is expressed in monetary units and is directly linked to the financial accounting system. It is important to note, however, that information do not have to be monetary in order to be material. A nonfinancial measurement can include such measurements as the amount of greenhouse gases produced, the amount of water consumed, or the share of products that will be eco-labelled. The interpretation and understanding of non-financial information may be more challenging than that of financial information since there may not be generally accepted reporting principles or the criteria may not be as authoritative as for financial information. Environmental audits may still be able to benefit from non-financial data despite these challenges. Accounting audits aim to determine whether, in all material respects, the financial statements adhere to an identified framework for financial reporting.

Box:1 AUDIT CASE STUDY: Coal Authority accounts of UK (2013-14)

After the United Kingdom's coal industry was privatized, the Coal Authority was established to oversee coal mining licensing and reclamation issues arising from coal and metal mining. Protecting the public and the environment, managing legacy properties, and maximizing value from and making information available are the core activities of the organization. According to the

Authority's annual report, the Authority has performed a number of tasks, including: - Developing and operating new mine water treatment schemes; - Investigating and repairing incidents affecting the public and subsidence;—Inspecting and monitoring closed mines' waste tips and entrances, while also overseeing activities that could impact coal sites, to guarantee the necessary risk assessment of gas, water, and ground stability is conducted.

IFRS requirements are followed when preparing the Authority's accounts, which comply with UK public reporting requirements. A revaluation of property, investments, equipment, plant and intangible assets is taken into account when preparing them under the historical cost convention. As a consequence, the accounts estimate future expenses for legacy liabilities that will last for 50-100 years. By the end of 2013-14, the provisions amounted to $1,066.0 million instead of $1,067.0 million. In the accounts, assumptions, trends, and changes to the discount rate are noted as influencing the provisions balance. At 31 March 2014, the Authority's accounts showed net liabilities of £975.7 million, which raised questions about its ability to continue to operate. According to the Report, the Authority expects to continue receiving Government funding in order to meet its liabilities for the foreseeable future, hence accounts have been prepared assuming the company will continue operating. It is the National Audit Office's responsibility to review the Authority's estimation of its provision in accordance with International Standards on Auditing (UK and Ireland) and obtain sufficient evidence to ensure that there are no material misstatements in the accounts. The accounting policy for the provision must also be assessed for appropriateness and adequate disclosure. Furthermore, the NAO should pinpoint any significant discrepancies in the audited financial statements by comparing the financial and non-financial data provided in the Annual Report.

8.7 Environmental Audit Process

EIA aims to anticipate problems before they arise. As part of the analysis, this information is provided to those making the decision on whether to approve or deny the project in order to predict its environmental impact. As a legal requirement in many countries, EIAs are also a part of many projects. Environmental Auditing is used to examine existing practices and assess the environmental effects of current activities, while EIAs are conducted when development has already been initiated and the project is already in operation. An overview of the project at that time gives an idea of what's happening.

8.7.1 Audit Initiation

An initiation letter is provided by audit program staff to the auditee several weeks before the audit for the following reasons:

- The audit should be informed to the organization;
- Create a communication process;
- Ensure that the audit will take place on the specified date and time;
- Develop an audit strategy, scope, and criteria;
- Describe the audit's topics; and
- Consider audit relief as an option.

Note: In addition to auditing information that has already been submitted to the ministry, the Compliance Audit Program can also audit information that has been previously submitted to the ministry without informing the auditees.

8.7.2 Audit Planning

After the audit is initiated, the lead auditor will prepare the audit plan and provide it to the auditee. The audit plan will outline the schedule for the audit, safety considerations and any other audit team requirements. Auditees may be asked to provide information about:

- Operating hours;
- Identifying key personnel and their contact information;
- Equipment necessary for personal protection;
- Training requirements in health and safety;
- A logistical consideration should be taken into account (e.g. directions to the site, meeting room location, food venue).

8.7.3 Conducting the Audit

The activities mentioned below are the part of audit conduction:

a. **Opening meeting**—The lead auditor leads the opening meeting, held at the start of the audit, with the objective.
    - A brief introduction to the audit team;
    - Analyze audit activities and confirm the audit plan;
    - Confirm communication channels and roles and responsibilities;
    - Ask the auditee any questions they may have
b. **Gather audit details and evaluate compliance**—As part of the audit process, the audit team reviews documents and records, tours the facility, and interviews

employees. Documents and records may be scanned or photos/videos taken by the audit team. Using the information collected, the audit team assesses compliance by the auditee.

c. **Daily Debriefs**—In order to identify potential non-compliance areas, the audit team will meet with auditee representatives at the end of each audit day. A report will be presented by the audit team that summarizes what was discovered that day and where additional information is required. To demonstrate compliance, the auditee presents information to the audit team during these informal discussions.
d. **Closing meeting**—After the audit is completed, the closing meeting is conducted to present initial findings and conclusions to the auditees, as well as explanations of the rationale behind them. Attendance at this meeting is encouraged for auditee organization personnel in charge of corrective actions and/or approving such actions.

8.8 Stages of Environmental Audit Process

Figure 8.1 illustrates how to prepare a corrective action plan after reporting audit findings. Assuring voluntary compliance is achieved by the lead auditor and the auditee is the task of the lead auditor. External and independent auditors conduct environmental compliance audits. Their interest in the outcome of the audit is neither financial, personal, or business. A site visit (if applicable) will be conducted after the auditor gathers all the relevant data. A report would be prepared after the facts were analyzed, the criteria were compared with the facts, conclusion drawn, and findings reported. Generally, there are three stages or phases to an Environmental Audit:

- Pre-Audit
- Audit
- Post-Audit

8.8.1 Stage-1: Pre-audit

In addition, the auditor will determine the scope and objectives of the audit, as well as which laws, policies, and procedures apply to the business's operations. In order to conduct an audit, an auditor must establish criteria. The best environmental management practices are based on company policies and procedures regarding environmental matters. Legislation and regulations are also important considerations.

- A team of auditors should be created that includes a variety of skills, talents, and perspectives
- Establish a plan for audits
- Documents to be requested and reviewed include:
  - Permits or applications for permits

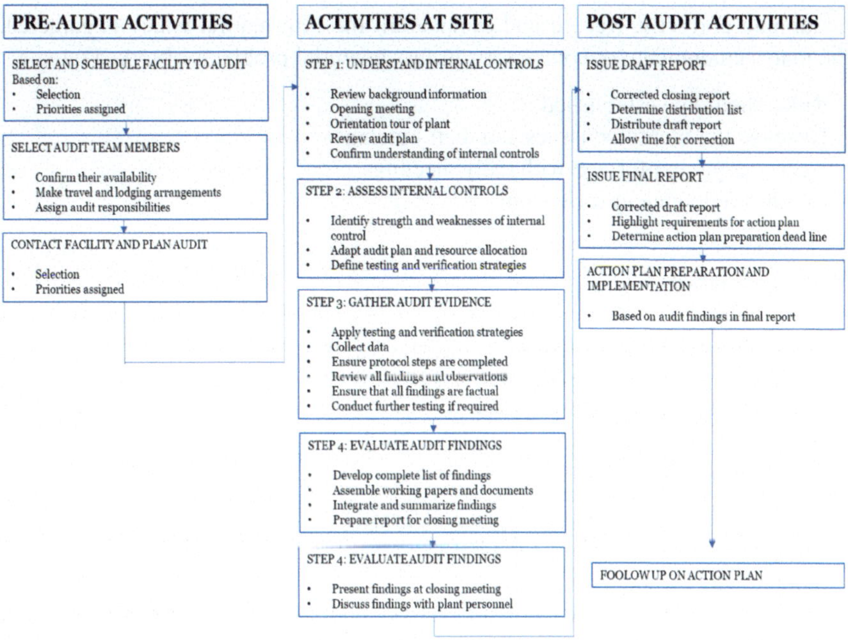

Fig. 8.1 Representing the basic steps of environmental auditing

- Records of production
- Detailed reports
- Audits conducted previously, including corrective actions taken and their status

- Establish a list of questions regulators might ask, followed up on prior audits, and requested additional materials

As violations are identified, begin filling out the Disclosure of Violations. Activities prior to the audit include:

1. Scheduling and selection of audit facilities,
2. Team selection for auditing,
3. Audit planning and contact with facility.

8.8.2 Stage-2: On-Site Audit (The Audit)

Auditor will assess if environmental regulations, policies, and procedures are being followed by the business. In addition to formal processes and practices, auditing requires understanding record-keeping and monitoring, inspection programs, as well as physical controls for preventing pollution. As part of the auditor's evaluation, he or she will test, observe, or ask questions to determine whether the controls are functioning correctly. Record and evaluate all the information which is gathered. The management team or contractors discuss potential problems observed on-site.

- Rules should be established
- Examine the identified issues and their outcomes
- Keep everyone informed with daily meetings
- Conduct a document review on:
  - Policies
  - Compliance
  - Training
  - Monitoring, control, and recording of air, water, waste, and noise
  - Procedures for responding to emergencies
  - Providing a complaint response
  - Review documents to ensure that they are complete, consistent, legal-compliant, and current.
- An inspection of the site should be conducted
- Compliance evaluation of operations
- Samples can be taken if necessary
- Verify that EHS policies are understood and applied consistently by EHS personnel, operations, management, and maintenance.
- Find out what concerns you

During the closure of meeting, all issues will be listed and discussed, and appropriate corrective actions should be developed for each issue. Site activities include the following steps:

1. Internal controls: a first understanding,
2. Internal control assessment,
3. Evidence collection for audits,
4. Audit findings evaluation,
5. The facility has been notified of the findings.

8.8.3 Stage-3: Post Audit

Once the on-site work is complete, a report will be prepared. A report on Environmental Compliance Audit will be compiled in which the auditor identifies whether the business is fully compliant, partially compliant, or non-compliant. There is a rundown of the company's policies and laws, as well as where the company complies and where it does not. Advisory action will be provided to improve overall compliance with environmental laws in the event of non-compliance. Whether a plan will be requested in the form of recommendations is up to the company. Environmental compliance audits conducted after a program has been established will refer to past reports to detect repeat noncompliance and assess implementation progress. Following actions are required for post audit:

- Formal preparation of Environmental Audit Reports and Disclosures of Violations
- Identify areas of concern and confirmed issues
- Follow-up actions and action items.

The following activities are performed after an audit:

1. A draft report will be produced
2. Final report production
3. Developing an action plan and implementing it
4. Keeping track of the action plan.

8.9 Documents Required for Environmental Audit

Before commencing an environmental audit, it is essential to collect and examine all necessary documents. Here are the six types of documents essential for initiating an environmental audit:

- Maps and floor plans of the facility are crucial for determining the audit's scope. They provide insights into any modifications made and procedures requiring a comprehensive assessment.
- Obtain copies of all environmental permits, as they are fundamental in confirming the facility's compliance with permit conditions.
- Collect environmental inspection reports and checklists as evidence of completed inspections as needed. Hazardous waste manifests, sampling data, and other reports demonstrate up-to-date recordkeeping and compliance, while also aiding in identifying potential compliance concerns.
- Maintain an inventory detailing all chemicals and raw materials used, their quantities, and locations to ascertain which environmental regulations are applicable to the facility.
- Environmental Plans and procedures, best management practices, and other relevant documents demonstrate how the facility addresses environmental concerns and avoids potential issues. This includes the EPA Risk Management Plan, Standard Operating Procedures (SOPs), and written emergency response programs, where necessary.
- Employee training records, assessments, and certifications indicate that employees are equipped with the necessary knowledge and skills to fulfil their responsibilities. These records also showcase the company's commitment to establishing and upholding environmental standards within the facility.

Summary

An environmental audit plays a crucial role in promoting sustainable development by holding organizations accountable for their practices. These audits assess procedures and identify necessary actions to ensure compliance with standards. Non-compliance can lead to significant consequences, such as expensive penalties. Environmental audits help cut down operational expenses by detecting problems early, reducing waste, and enabling the company to strategize corrective measures. Environmental auditing should not be mistaken for environmental impact assessment (EIA). Even though they are both tools for environmental management and use similar terms such as impact, effect, and significance, there are distinctions between the two. Environmental auditing encompasses a range of management practices utilized to assess a company's environmental performance. It involves rigorous inspection of systems and procedures against established standards or regulations, which are employed to collect and analyze environmental data during an environmental review.

Questions/Exercise

1. Describe environmental audit, its objectives and key features.
2. How environmental audits benefits the industries.
3. Write an essay on environmental monitoring.
4. Write in detail about various stages of environmental audit.
5. Enlist various documents required during environmental auditing.

References

https://www.nrep.org/blog/environmental-audit

https://sitemate.com/uk/resources/articles/environmental/environmental-auditing-process/

https://www.envass.co.za/environmental-audit-types-and-benefits/

https://www.routledge.com/Environmental-Performance-Auditing-in-the-Public-Sector-Enabling-Sustainable/Prasad/p/book/9780367856045

https://static1.squarespace.com/static/57019a6db6aa607cbb909ab2/t/5923b3d12e69cf8a531eded1/1495512024499/ISSAI+5120+E+Endorsement+version_2016.pdf

Kells S, Hodge G (2009) Performance auditing in the public sector: reconceptualising the task. J Contemp Issues Bus Gov 15(2):33–60

Pollitt C, Summa H (1999) Performance audit and public management reform. In: Performance or compliance, pp 1–10

Chapter 9
Environmental Management System (EMS)

An Environmental Management System (EMS) allows an organization to recognize its environmental aspects and identify those with a notable impact on the environment. By doing so, the organization gains insight into its environmental interactions. This insight aids in pinpointing areas requiring environmental enhancements or controls and establishing priorities to enhance environmental performance. An organization can create an environmental management system in alignment with the ISO 14001 standard as a key component of its environmental strategy and to meet regulatory requirements. The EMS directs efforts towards fulfilling the objectives outlined in the organization's policy, such as minimizing adverse environmental effects of its offerings and enhancing their positive impacts. International Standards for environmental management offer strategic tools for addressing environmental concerns. ISO 14001, which outlines the requirements for an environmental management system is one of the most recognized standards. Along with a series of companion documents, it constitutes the ISO 14000 standard family.

9.1 Environmental Management System

An Environmental Management System comprises processes and practices that help an organization lessen its environmental effects and enhance operational efficiency. EPA continually review the progress to develop and maintain the environmental management system at every facility operation, laboratories and offices that primarily focuses on the reduction of agency's environmental footprint. Reviewing and assessing consistently aids an organization in enhancing its environmental performance. The EMS serves as the framework that guides an organization towards achieving its environmental objectives. Regularly reviewing and assessing an organization's environmental performance can help identify areas for enhancement and implementation. Each organization customizes its EMS to align with its specific

© The Author(s), under exclusive license to Springer Nature Switzerland AG 2024

R. Bhateria et al., *Environmental Impact Assessment*,
https://doi.org/10.1007/978-3-031-66797-8_9

goals and objectives, without specifying a predetermined level of environmental performance that must be reached.

Organizations can address their regulatory requirements in a systematic, cost-effective manner by implementing an EMS. By taking this proactive approach, non-compliance risks can be minimized, and health and safety practices can be improved. An EMS can enhance operational control and employee stewardship, while also tackling non-regulated concerns like conservation of energy. An EMS consists of the following basic elements:

- Analyzing the environmental goals of the organization.
- Assessing the effects of the project on the environment, as well as the compliance obligations (legislative and additional requirements).
- Reducing impacts on environment and conforming to compliance obligations by setting environmental goals and aims.
- Developing agendas to reach these objectives and targets.
- Achieving objectives by tracking and assessing progress.
- Improving the EMS by reviewing its progress.

A management system consists of three primary processes:

1. Identifying significant environmental aspects and impacts of the core processes and their outputs
2. Assuring employees are competent, providing infrastructure, distributing EMS information, monitoring, and evaluating environmental performance are all important supporting processes.
3. Controlling documents, maintaining records, and conducting internal audits are all processes supported by a management system

9.2 Purpose of an EMS

As part of an Environmental Management System, regulations are implemented along with targets and goals to ensure that the organization complies with the policies. It is essential for a business to identify various environmental aspects so that EMS can articulate appropriate measures effectively. It is important to understand legal requirements and reference company goals when planning appropriately. An EMS makes it simple and affordable for companies to meet their regulatory requirements. Health and safety practices can be promoted for workers and the general public through it and minimize the risk of non-compliance. Environment Management Systems include the following elements.

- Performing the necessary improvements to an EMS on a regular basis and tracking its progress
- Assessing the impact of the company's operations on the environment as well as its compliance obligations
- Complying with legal obligations with regard to environmental goals and targets

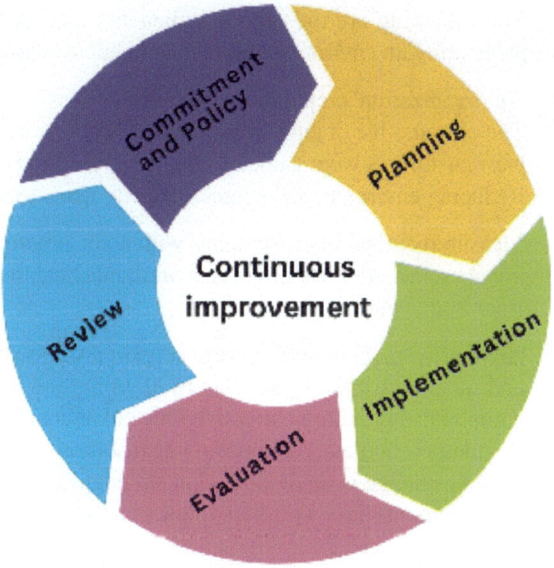

Fig. 9.1 The continuous improvement cycle

- Engage employees and customers in environmental awareness
- Assessing progress towards achieving environmental goals

An organization's continuous improvement of environmental performance is one of the objectives of an EMS (Fig. 9.1). Repeating cycles are part of the system. Organizations commit to an environmental policy, then establish a plan to improve environmental performance using the policy as a guide. Implementation is the next step. Once the objectives and targets have been achieved, the organization evaluates its environmental performance. It is necessary to take corrective action if targets are not being met. Following this evaluation, top management reviews the results to see if the EMS is functioning properly. Revised targets are established in a new plan following a review of the environmental policy by management. The company then puts the revised plan into action, leading to ongoing enhancements as the process continues.

9.3 Benefits of an EMS

Environmental management systems have the following advantages:

- Environmental impacts must be considered holistically
- A focus on only critical processes and aspects
- Achieving global recognition through the application of tried-and-true sophisticated methods
- Establishing constructive connections with regulatory bodies

The following are the economic benefits that an organization can expect from implementing an environmental management system:

- An organization's reputation and image
- The cost and fee of environmental protection is lower
- New customers were gained through increased access
- Reducing environmental sources directly leads to savings

Measurement of improvements within an organization can take one of three approaches, in addition to obvious quantifiable benefits such as reduction of pollutant emissions and waste.

- An improvement in management support processes is accomplished in qualitative and quantitative ways, such as by training and educating employees, ensuring compliance, or taking corrective/preventive action.
- Regulatory bodies, community organizations, or any stakeholders perceive a positive change in an organization's reputation or relationships.
- Improvements have a positive impact on the bottom line by allowing cost savings or cost avoidance

Due to the non-prescriptive nature of ISO 14001, environmental management systems can be tailored to meet the needs of any organization. Identifying the organizational maturity level and taking necessary actions to advance to the next stage is the best way to realize the full potential of an environmental management system when one organization fails to achieve the expected benefits from its management system. Figure 9.2 showing the evolutionary development of EMS. Following are the internal and external benefits:

Internal

- Most organizations spend the majority of their EMS resources on staff and managers.

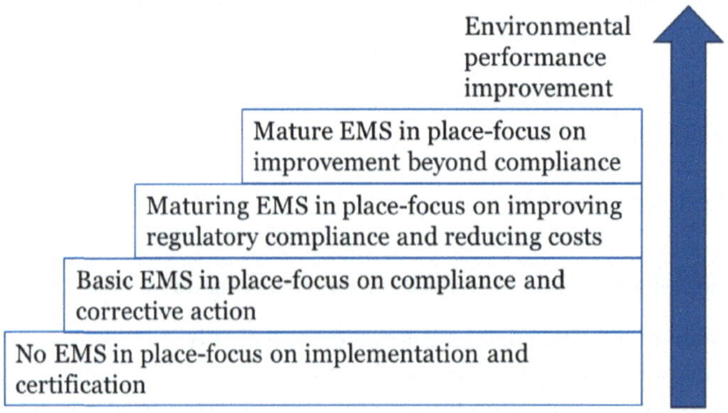

Fig. 9.2 Depicting development in environmental management system through evolutionary model

- Employees' other time

External

- Assistive consulting
- Employees are trained outside the organization

Potential Benefits

- Performance improvement in terms of the environment
- Compliance enhancements
- Prevention of pollution
- Conservation of resources
- Customer/market expansion
- Cost reduction/improved efficiency
- Boosted employee morale
- Investors, lenders, regulators, and public perceive the company in a more positive light
- Awareness and responsibility of employees toward environmental issues

9.4 A Basic EMS Framework—PDCA Cycle

ISO 14001 is a globally acknowledged standard for environmental management systems, created by the International Organization for Standardization (ISO). Over 300,000 organizations across 171 countries, including the United States, have obtained ISO 14001 certification. The standard underwent revision in 2015. In order to determine the needs of current and potential users, ISO conducted a continuous improvement survey. ISO 14001 outlines general requirements and guidelines to ensure that implementing the system will not negatively affect the environment and may enhance environmental performance. A nonprescriptive standard such as ISO 14001 explains what should be done, not necessarily how. ISO 14001 is based on the plan-do-check-act (PDCA) model of improvement, which must be applied consistently for the standards to be upheld and the benefits realized. EMS that adhere to ISO 14001 consist of the following primary operational components:

1. *Plan:*
    - Aspects of the environment
    - Requirements legal and otherwise
    - Programs, objectives, and targets
2. *Do:*
    - Authority, resources, and responsibilities
    - Awareness, competence, and training
    - Communicating
    - The documentation

- Documentation control
- Controlling operations
- Response and preparation for emergencies

3. *Check:*

- Measuring and monitoring
- Compliance evaluation
- Corrective and preventive actions for nonconformity
- Keeping records under control
- Conducting internal audits

4. *Act:*

- Review of management

9.5 EMS Standards: ISO 14000 Series

During the development of ISO 14000, the key principles taken into account were: they should contribute to enhancing environmental management; they should be universally applicable across all nations, however, it is important to prioritize the interests of the public and the standards users. In order to meet the differing needs of organizations of any size around the world, they should be cost-effective, non-prescriptive, and flexible. It is important that they are flexible enough to be verified either internally or externally, as part of their flexibility. Ideally, they should be based on scientific evidence.

The key aspect is their functionality, practicality, and user-friendliness. The initial ISO 14000 series standard was released in September 1996. ISO 14001 refers to the standard for EMS. Additionally, a guidance document was issued alongside ISO 14001, known as ISO 14004. For auditing the EMS, ISO 14010, 14011, and 14012 were introduced in October 1996. An Environmental Labelling (EL) standard was published in 1998 and a declaration standard was published in 2000. A company's environmental management system aims to implement eco-friendly practices within the organization, enabling the production and sale of products and services with reduced environmental footprints, thus reaping the rewards of sustainable operations. Basically, ISO 14000 series documents can be divided into two groups: "organization-oriented and product-oriented". As an organization-oriented standard, it provides complete guidance for the establishment, maintenance, and evaluation of an EMS. Additionally, it relates to environmental functions and systems across the entire organization. In addition to environmental systems and functions, it involves other aspects of organization-wide operations.

Here is a list of ISO 14000 standards, TRs, and guides focused on organizations:

1. **ISO 14001:2004,** Guidance for Use of the Environmental Management System Specification

2. **ISO 14004:2004,** Principles, systems, and supporting techniques for environmental management systems
3. **ISO 14010:1996,** Environmental Auditing Guidelines (General Principles)
4. **ISO 14011:1996,** Environmental Auditing Guidelines—Includes environmental auditing procedures, environmental management audits
5. **ISO 14012:1996,** Audit Guidelines for Environmental Auditing—Guidelines for Environmental Auditing, Environmental Auditors Qualification Criteria
6. **ISO 14031:1999,** Guidelines for evaluating environmental performance
7. **ISO/TR 14032:1999,** A framework for environmental performance evaluation (EPE) in environmental management
8. **ISO/TR 14061:1998,** Guidelines for Forestry Organizations on the use of ISO 14000 and ISO 14001 EMS.

It consists of a set of standards related to environmental management that are designed to help organizations manage the environment. These are to:

- Reduce the environmental impact of their operations
- Abide by laws, regulations, and other requirements relating to the environment
- Continuously improve in the above areas

One of the most important standards of the ISO 14000 family (Box 1) is ISO 14000 which represents the core set of specifications used by organizations to create and implement effective environmental management systems (EMS) (Fig. 9.3). ISO 14000 is an internationally agreed standard that must be followed by an environmental management system. Using resources more efficiently and reducing waste can give organizations a competitive advantage and gain stakeholder trust. In accordance with the ISO series, environmental management includes "Environmental Management Systems (EMS), Environmental Auditing & Related Investigations (EA&RI), Environmental Labels and Declarations (EL), Environmental Performance Evaluation (EPE), Life Cycle Assessment (LCA), Terms and Definitions (T&D)".

Box 1: ISO 14000 series

- ISO 14000—EMS—General Guidelines on Principles, Systems and Supporting Techniques
- ISO 14001—EMS—Specification with Guidance for Use
- ISO 14004—EMS—General Guidelines on Systems, Principles and Supporting Techniques
- ISO 14010—EA—General Principles of Environmental Auditing
- ISO 14011—EA—Auditing of Environmental Management Systems
- ISO 14012—EA—Qualification Criteria for Environmental Auditors
- ISO 14013—Management of Environmental Audit Programs
- ISO 14014—Initial Reviews
- ISO 14015—Environmental Site Assessments
- ISO 14020—EL—Basic Principles of Environmental Labeling

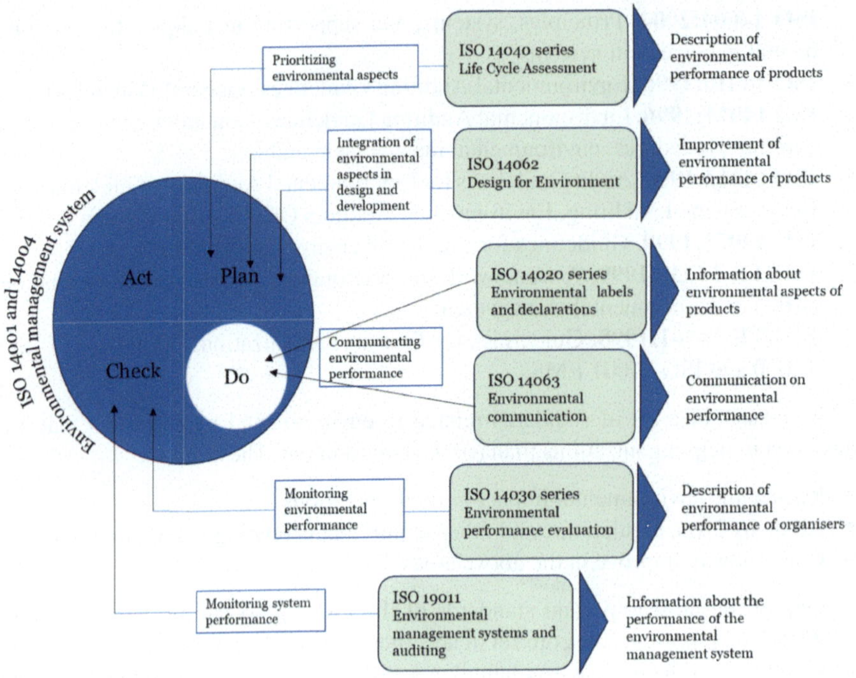

Fig. 9.3 Representing relationship between ISO 14000 series standards and their application (*Source* ISO 2000)

- ISO 14021—EL—Self Declaration—Environmental Claims—Terms and Definitions
- ISO 14022—EL—Symbols
- ISO 14023—EL—Testing and Verification Methodologies
- ISO 14024—EL—Practitioner Programs—Guiding principles, practices and certification procedures of multiple criteria (type1)
- ISO 14031—Environmental Performance Evaluation
- ISO 14040—LCA—General Principles and Practices
- ISO 14041—LCA—Goal and Definition/Scope and Inventory Assessment
- ISO 14042—LCA—Impact Assessment
- ISO 14043—LCA—Improvement Assessment
- ISO 14050—Terms and Definitions
- ISO 14060—Guide for the Inclusion of Environmental Aspects in Product Standards

The ISO developed the 14001 standards as the basis of its EMS framework. Based on the PDCA methodology, this framework has been an official international standard since 1996. In accordance with ISO 14001, an EMS is divided into five main stages:

1. *Commitment and Policy*—Management of the organization sets an environmental policy that focuses on improving the environment. In order for the EMS to function, the policy must be in place.
2. *Planning*—As part of its planning process, organizations identify environmental factors that affect the operations they conduct. Items that can have adverse effects on people and/or the environment, like air pollutants and hazardous waste, are known as environmental aspects. By identifying criteria that are deemed crucial by the organization, it determines the significant aspects. Health and safety of workers, environmental compliance, and cost may be among the criteria an organization chooses to evaluate. An organization sets targets and objectives after determining significant environmental aspects. The objective of an environmental strategy is to minimize the use of chemicals (e.g., use of chemical X as little as possible). As a result of the objectives, targets are detailed, quantified requirements (e.g., reduce the use of chemical X by 25% by September 2030). Meeting the targets requires a detailed action plan as the final stage of the planning process. To accomplish this, responsibilities will be assigned, a schedule set, and clearly defined steps will be outlined.
3. *Implementation*—During implementation, an organization utilizes the resources necessary for carrying out the action plan, including human resources, financial resources, and others. During the implementation stage, employees must be trained and aware of what is expected of them (including interns, contractors, etc.). Furthermore, the implementation stage includes documentation, adherence to operating procedures, and establishing communication channels both internally and externally.
4. *Evaluation*—Monitoring how well a company is doing at meeting its objectives and targets is part of the company's evaluation process. Corrective measures are taken if necessary.
5. *Review*—Top management reviews the evaluation results to assess the implementation of the EMS according to the plan. They evaluate the environmental policy to ensure alignment with the organization's values. A revision is then made to the plan to optimize its effectiveness. A company's continuous improvement journey begins with the review stage.

9.5.1 Principle and Structure

In the context of British standards, this regulation is known as BS EN ISO 14001:2015, while in European standards, it is identified as EN ISO 14001:2015. A certification scheme for environmental management standards is ISO 14001. A "process approach" recognized in quality management systems such as ISO 9001 is incorporated in ISO 14001 as a High-Level Structure (HLS). All ISO management systems are drafted using the HLS, which offers a common approach to integrating management systems.

In addition to organizations in the industrial sector, ISO 14001 applies to all types and sizes of businesses. It not only outlines EMS requirements but also covers environmental aspects and their associated impacts that organizations have the ability to manage or influence. This helps in the implementation of an environmental policy and the setting of objectives. It is based on the principle of continuous improvement in environmental performance that is an integral part of ISO 14001. Another international standard in the ISO 14000 series provides guidance on principles and practices of environmental management. An organization must commit to embracing the following within its policies, objectives, and targets:

- The pursuit of continuous improvement
- Pollution prevention
- Observance of applicable environmental laws and regulations.

An accredited third party can certify a facility against ISO 14001 as a specification standard. A self-declaration option is also provided in the standard. In order to maintain ISO 14001 certification, the EMS must be reviewed periodically.

9.5.2 Why ISO 14000?

Organizations must establish effective environmental management systems in order to comply with ISO 14000. Standardization aims to offer cost-effective tools for structuring and implementing environmental management information through best practices. Industry stakeholders acknowledged the necessity for standardization, leading to the creation of the ISO 14000 family to fulfill this requirement. Prior to this, comparing and cooperating on environmental management systems proved challenging due to varying organizational methods.

ISO 14000 is a crucial part of today's global economy because it addresses the economy, the environment, and the social side, or the "triple-bottom line." There are a number of fields involving interdisciplinary collaboration that can benefit from ISO 14000, including finance, insurance, marketing, regulation, and many others. Standards promote a proactive rather than a reactive approach to environmental impact, and even address unregulated factors like energy or resource allocation, helping organizations reduce their environmental impact (Box 2). Among the most important reasons for implementing ISO 14000 are:

Risk reduction—Energy costs are on the rise and supply is uncertain, so companies can minimize risk by assessing and recognizing these factors.

Leading by example—In order to comply with government regulations and avoid fines, companies typically adopt sustainability strategies. It is now seen as a progressive environmental policy trend to embrace sustainable environmental practices.

Tax incentives—Companies that embrace and follow environmentally friendly standard operating procedures can qualify for a range of tax incentives provided by state and local governments. Investment tax credits, production tax credits, and consumption tax credits can be used as incentives. We can also take advantage of cash rebates, tax exemptions, and improved capital expenditures.

For employees—According to "World Resource Institute", many employees are excited and motivated by the possibility of contributing to bigger causes at their workplaces. Employees tend to stay engaged and productive when environmental policies are integrated into the corporate culture.

Brand image and Public Relations (PR)—An ISO 14000-compliant environmental policy can go a long way towards improving customer relations and ultimately boosting brand perception over time.

Renewable & non-renewable resources—Fossil fuels and renewable resources are limited and should be handled carefully by following sustainability-focused standards. Resource scarcity will lead to an increase in cost. Sustainability plans or alternative plans will have to be adopted by companies in anticipation of this eventuality.

Competition—Around two-third of nearly 3,000 executives surveyed by MIT in 2011 believed that sustainability is essential to compete effectively today.

Box: 2 Findings of a study by Unilever

According to 33% of consumers, brands that are socially or environmentally responsible are more likely to be purchased. Research shows that 21% of respondents are inclined to buy from brands that prioritize sustainable packaging. Consumers highly value knowing that their preferred brands are environmentally conscious. Given the current concerns about global waste and limited resources, consumers are keen to understand what measures companies are taking to promote environmental sustainability, including land conservation, animal welfare, water quality management, and waste disposal practices. The consumer wants more from quality and affordability than just those two factors. Nowadays, stakeholders are closely and effortlessly interested in a company's environmental and social responsibilities.

Companies must establish, execute, and convey environmental management and sustainability principles that match consumer and stakeholder expectations. This is crucial to address the significant spotlight and accountability associated with social and environmental responsibility. Managing your company's environment can make more sense economically in the long run than simply being eco-friendly. It is possible for businesses to reduce energy costs, gas costs, and other costs by applying principles of environmental management.

9.5.3 History of ISO 14000 Series

Conventional quality methods have concentrated on enhancing efficiency, profitability, and customer satisfaction through the reduction of defects and waste. Quality initiatives have recently begun to incorporate environmental and sustainability goals. Because environmental practices are typically viewed as incurring extra expenses in the business arena, it is crucial for quality practitioners to pinpoint areas where environmental and sustainability concerns, like waste reduction and renewable energy usage, align with the interests of both businesses and their customers. Compliance management evolves from compliance to sustainability in the following ways:

- ***Between 1990 and 1994***, Compliance Management (CM) centered around regulations and allowed environmental departments to handle issues that arose.
- ***From 1994 to 1998***: Environmental management systems (EMS) improved environmental management in organizations.
- ***1998 to 2002***: The use of web-based systems and the integration of multiple systems have become commonplace in environmental information management systems (EIMS).
- ***In 2002 to 2006***, environmental process management systems (EPMS) were primarily used to drive improvements through the use of quality tools.
- ***2006–current***: To achieve sustainability, it is critical to adopt best practices to address risk and uncertainty, as well as integrating environmental, social, and economic goals.
- It was published by the BSI group in 1992, and it is considered one of the first environmental management standards.
- A family of standards called ISO 14000 was defined by the International Organization for Standardization (ISO) in 1996.
- 2004 marked the revision of ISO 14001 standards.
- As of September 2015, ISO 14001 has undergone a revision (Fig. 9.4).

9.5.4 Organization Standards

Organizational standards are an investment that can be viewed as establishing the essential foundations for an environmental management system. This can be accomplished by implementing ISO 14001. A company must use the auditing tools in the ISO 14010 series in order to properly implement ISO 14001 in the company. Additionally, the ISO 14030 series stipulates parameters for evaluating an organization's environmental performance. The organization can then take advantage of its EMS once these systems and tools have been implemented.

Fig. 9.4 Timeline showing history of ISO 14000

9.5.4.1 EMS

EMS is applicable to organizations of all types and sizes. ISO 14001:2015 outlines the requirements and gives guidance on how they should be implemented. Implementing environmental management includes defining policies and objectives, preventing pollution, and promoting continuous improvement, among other aspects. A competent body must audit the EMS for full compliance to lead to certification. Several environmental guidelines for ISO 14001 can support the implementation process, including standards in the ISO 14000 series.

The European environmental management system, known as EMAS, serves as a tool for assessing, reporting, and enhancing the environmental performance of companies and other entities. A similar Environmental Management System (EMS) is a mandatory part of ISO 14001 certification. For EMAS acceptance, the EMS segment must hold ISO 14001 certification. Alongside a preliminary assessment, obtaining EMAS registration necessitates a comprehensive environmental report in simple language and an independent verification. It also offers direction on the gradual adoption of environmental management systems, encompassing performance metrics, for various organizations, including SMEs. In order to meet an organization's objectives, ISO 14005 allows it to concentrate on specific environmental aspects. A phased approach to EMS implementation was also provided by BSI's EMS guidance standard, BS8555, which has now been replaced with ISO 14005.

9.5.4.2 Environmental Auditing

EMS are the basis for auditing. An audit can be defined as a comparison between the current state and the anticipated state, along with an assessment to verify if the current state aligns with the expected conditions. In the same way, financial audits, quality assurance, regulatory compliance audits, and systems audits all use this philosophy. Understanding why audits differ begins with understanding the common elements. Audit components come in several forms, and they are all defined differently. This terminology is defined in ISO 14010 for EMS audits, but it can also be used in other audits. Although compliance audits were originally considered by the ISO committee, they were not developed.

Since the concepts and processes outlined in 14011 were suitable for various audits beyond EMS audits. Analyzing evidence and comparing them with criteria is at the core of an audit. By conducting interviews, observing, and reviewing documents, the evidence is gathered objectively. Audit criteria define what conditions should be based on expectations or rules. An audit is differentiated by its criteria. For example, regulatory compliance audits use regulations as their criteria. In the case of an EMS audit, the criteria would be the essential elements that need to be incorporated into the system. As a result, ISO 14001, the specification standard, would be the appropriate EMS criteria. A condensed version of the EMS audit incorporates the following general characteristics that are essential to any audit, i.e.

- They are systematic and have preplanned approach instead of being random
- Inquiry and critical consideration should accompany them—They are free from bias or prejudice
- Environment-related activities and results are of concern to them
- In accordance with planned arrangements, make sure that these activities are executed efficiently and consistently.

Why Perform EMS Audits?

As part of ongoing monitoring and measurement, some type of monitoring activity is required to confirm that the defined EMS system functions effectively. Any such additional monitoring can potentially have the following benefits based on EMS audits.

EMS Policies are used to verify that they are understood and followed, as:

- The system is being implemented according to instructions, which gives management confidence in its effectiveness. Through them, deficiencies can be identified, corrective measures can be agreed upon, and effectiveness can be verified.
- In order to prevent environmental problems from occurring, they provide early warnings of system weaknesses.
- Their convenience facilitates the investigation of operations in a particular area, including environmental issues.

- Again, when personnel from different parts of the organization are involved, the opportunity to exchange ideas is created, allowing successful features from one region to be applied to another, if possible.
- Employees' commitment and motivation can be increased by involving them more widely in business operations.

Plan, execute, and report are the three major parts of an audit. As part of ISO 14011, the audit procedures are described, and suggestions are provided for setting up an EMS audit program. An organization must establish auditing programs and procedures in accordance with ISO 14001 regulations. A very important step in any audit is planning the audit. The same applies to EMS audits. This includes creating a detailed audit plan, assigning team roles, determining necessary documents, and handling any exceptional circumstances that may arise. Imagine going on vacation without planning; you won't know where you will go, what you will do, or how long you will be away. That's why planning is important.

The Audit Plan

The audit plan outlines the audit's scope, objectives, criteria, and schedule. It details the areas to be audited, the auditors, and the timing of the audit. The document also specifies the individuals responsible for conducting the audit. Moreover, the plan may incorporate guidelines on report formatting, distribution, and meeting procedures. This audit plan mirrors the ISO 14001 EMS audit process in terms of procedures, methodology, and program content. For instance, the plan may schedule four partial audits per year for the entire EMS. A procedure is then created based on this schedule. Based on ISO 14011, the following elements should be included in an audit plan:

- Scope and objectives of the audit;
- Criteria for auditing;
- Organization and functional units of the auditee to be audited;
- A description of the functions and individuals that have significant direct responsibility for the auditee's EMS within the auditee's organization;
- An auditee's EMS should be identified as having high audit priorities;
- Audit procedures appropriate to the auditee's organization for auditing the EMS elements;
- Languages used for audit reporting and working;
- Reference documents shall be identified;
- Major audit activities' expected duration and time;
- Audit dates and locations;
- Identifying members of the audit team;
- Meeting schedules with auditee management;
- Keeping confidential information;
- The format and content of the audit report, as well as the expected date on which the audit report will be released;
- The length of time documents must be retained.

Roles and Responsibilities in Audit Conduct

With the audit plan prepared, the audit team assigned, and working documents defined, the audit is ready to be conducted. Information is gathered and compared to criteria to evaluate the level of compliance with planned arrangements. The lead auditor ensures the audit scope and plan are adhered to in order for the audit to be efficient and effective.

Auditor—Auditors should be responsible for the following activities and responsibilities:

- Follow the lead auditor's directions and provide support;
- Assuring that the audit is carried out objectively, effectively, and efficiently;
- Determining audit findings and conclusions about the EMS involves gathering and analyzing appropriate and adequate audit evidence.
- In accordance with the lead auditor's instructions, prepare working documents;
- Audit findings must be documented;
- As required, safeguard audit documents and return them;
- Writing the audit report with the assistance of the auditor.

The auditing team—Audit team members should be selected based on their experience and expertise so that the audit process can be conducted efficiently. Accountable auditees should be responsible for the following:

- As necessary, providing employees with information about the scope and objectives of the audit;
- Assuring that the audit team has the facilities needed to carry out the audit effectively and efficiently; Performing the audit Collecting Evidence Once the audit scope has been agreed upon with the auditee and the client, it is time to review the accompanying documentation, which generally consists of:
  - For the audit area, the audit manual and procedures must be provided
  - Documentation and specifications pertaining to the audit area
  - Any audit checklists related to that area and the findings of the last audit
  - Documents pertaining to corrective action analyses

Checklists—During the discussion of audit preparation earlier, checklist preparation was highly emphasized. It is important to keep these checklists available throughout the audit. Ideally, you should not follow them blindly, but rather use them to ensure you have covered all relevant topics. Spaces incorporated into the checklist may be useful for recording audit information in some cases. For objective evidence of effective implementation, internal audit procedures often require the checklist to be retained. Auditing's most sensitive part, interviewing, is probably one of the most difficult things to teach, and more of an acquired skill. A series of questions is used to gather information from another individual during an interview. The type of question can prompt different kinds of responses, but this may sound simple at first.

Generally, closed questions (yes–no answers) do not yield detailed explanations. Closed questions are not practical for measuring someone's understanding of a concept. The auditor may also close questions when it is time-sensitive or when there is no time available. Also, keep in mind that the auditor can clarify a point with additional questions. EMS description has been clearly outlined; therefore, the documented EMS system encompasses formal documentation of policies, procedures, organizational structures, and responsibilities. Typically, environmental or departmental procedures and work instructions are incorporated into an EMS manual. Here are some key reasons why a documented system is crucial in facilitating the Audit Process. After gathering evidence, the audit team convenes to determine the next steps. To determine if the EMS conforms to plan, evidence is compared to criteria. As a result, the results of the assessment will either be positive or negative.

9.5.4.3 Environmental Performance Evaluation

ISO 14031 defines environmental performance evaluation (EPE) as a, "reliable and verifiable internal process and management tool designed to provide management with reliable and verifiable information that enables them to determine whether their environmental performance meets the criteria established by their management." Plan, do, check, and act is the final stage of the model (Jash 2000). A given point in time can be assessed by the results of the EPEs (including EPIs and ECIs). Examples of how EPE can be applied in practice are provided in ISO 14031. It is possible, for example, to identify the relative importance of environmental aspects based on:

- A measure of the relative importance of various environmental aspects concerning services, products, and activities, as well as the potential impacts they might have on the environment.
- Analyzing the impact that activities, products, and services may have on environmental conditions using information about the condition of the environment.
- Inputs, discharges, waste, emissions, and risks associated with material and energy inputs, discharges, waste, and emissions are analyzed and interpreted by the organization.
- The organization's significant environmental aspects are determined by identifying and analyzing the views of interested parties.
- A list of organizations' activities that may have been collected by the organization in compliance with environmental regulations or other requirements.
- The organization's products should be designed, developed, manufactured, distributed, serviced, used, re-used, recycled, and disposed of according to their environmental impact.
- In addition to identifying outsourced activities or processes, an organization should identify which activities have significant environmental impacts.

An EPE's main benefit is that it identifies areas that can be improved, whether within the management system, the organization's activities, or both. It is crucial to consider several questions such as:

- Do you observe a positive trend in your organization's environmental performance over time?
- Does the EMS process have a high level of cost-effectiveness?
- Is the correct emphasis being placed on the right aspects and impacts?
- Does the data source and analytical method match the purpose of the research?

A relevant, reliable, measurable, and comparable set of data should be collected and used in the EPE system. In the context of the overall review process and the plan, do, this information can be fed back into the EMS. It is possible to restart the check, act methodology in order to ensure that environmental performance is continuously improved.

9.5.5 Product-Oriented Standards

An organization that adopts product-oriented standards generates revenue rather than investing in or incurring costs of standards that are oriented toward the organization. In order to improve the sustainability of its products and services, companies must first analyze and evaluate environmental problems and weaknesses throughout their entire life cycle. The ISO 14040 series of LCA tools are necessary for this. Products and services must be improved by correcting identified environmental weaknesses. An ISO/TR 14062, "Design for the Environment," design tool will be necessary to accomplish this. Following ISO 14020, the corporation affixes environmental labels or makes environmental claims about the products before they are launched into the market. In this way, consumers are more likely to buy the products and services, which will, hopefully, lead to an increase in sales and market share (Rocha and Silvester 2001).

There were two types of ISO 14000 standards published by ISO in response to rising demands for preserving the environment, conserving resources, and promoting a sustainable society: management oriented and product oriented (Fig. 9.5). Apart from the ISO 14000 product-oriented standards, ISO 14000 also includes guidelines for environmental labels, environmental declarations, and life cycle assessments. In addition to providing guidelines for improving the environment performance of products and services, these standards are intended to help assess how products and services perform environmentally. The standards can thus be used as guidelines in international trade for taking environmental consideration into account. Standardization has the capability of becoming both a technical facilitator and a trade barrier.

9.5.5.1 Environmental Labeling

A product or service's environmental label indicates whether it is recyclable, or contains noxious ingredients, as well as its overall environmental benefits. The

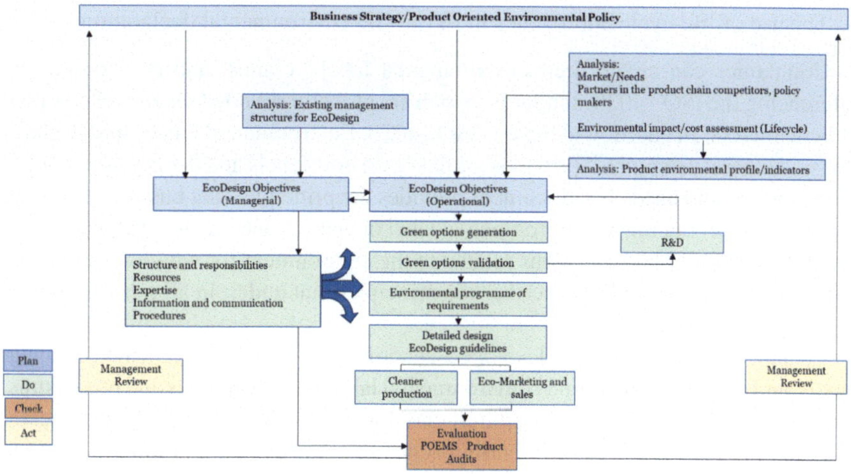

Fig. 9.5 Theory to practice product-oriented EMS in Europe

presence of this information can have a positive impact on consumers' decisions, especially when they are increasingly inclined to buy eco-friendly products. This makes for a win–win situation for all. In recent years, 'greenwashing' and exaggerated marketing claims have become more common due to the proliferation of such labeling. A credible and level playing field can only be achieved through internationally agreed-upon and harmonized criteria and labeling methods. A fundamental role is played by ISO International Standards here. Within technical committees (TCs), ISO standards are developed by groups of experts. ISO and its members convene TCs from around the world consisting of industry representatives, non-governmental organizations, government officials and other stakeholders. There are different TCs that focus on different subjects or areas of expertise. It serves as the central point for establishing standards for environmental labeling and operates as a subcommittee of ISO/TC 207, focusing on Environmental Management. A total of six standards have been published so far by the group, which is composed of experts from 78 countries.

Businesses are able to create their own environmental labels based on industry ISO standards in a way that communicates the value of their products and builds customer loyalty and trust. The consumer benefits when businesses disclose accurate, clear, and comparable information about products and services based on ISO standards. Moreover, ISO standards can assist regulators with the establishment and measurement of criteria for environmentally related incentive schemes, enabling them to create public policies that meet international commitments. The many challenges brought on by climate change can also be addressed through ISO standards.

- There are clearly defined criteria for products that qualify for eco-labelling schemes under "Type I" environmental labeling
- Products and services claiming a "Type II" environmental claim without criteria or labeling

- The use of life-cycle approaches in "Type III" environmental declarations

Companies can create their environmental labels, claims, and declarations by referencing the ISO 14020 standards, which are globally acknowledged and accepted as benchmarks. In addition to these, "ISO 14020, Environmental labels, and declarations—General principles" provides guidance on how labels and declarations should be developed and used. The document provides the principles that businesses should follow when producing an environmental label or claim to ensure their information is accurate and credible. For claims related to the environment for which no standards have yet been developed, it describes the principles that underpin all other standards in the 14020 series.

Ecolabelling schemes are developed according to ISO 14024, "Environmental labels, and declarations—Type I environmental labeling—Principles and procedures, a document that describes the requirements for developing "Type I environmental labelling programmes", or ecolabelling schemes" As soon as a product or service meets certain criteria, a mark or logo is awarded to the product or service. According to ISO 14024, product categories are selected using product-environmental criteria, product-function characteristics are assessed, and compliance is demonstrated. As well as establishing the certification procedure, it also specifies how the label will be awarded. Global Ecolabelling Network (GEN), an international federation of ecolabelling organizations, has adopted the ISO 14024 standard as a benchmark for environmental labels.

9.5.5.2 Life-Cycle Assessment

The life cycle data declarations allow comparisons between similar products by presenting quantified environmental information. Regulatory requirements and business-to-business communications can also be facilitated through them. As the "nutrition labels for the environment", they are commonly referred to as Environmental Product Data (EPD) sheets. A product or service's environmental impact can be assessed and compared using information and data provided by such labels. It details the principles and methods for creating data for these declarations, along with the criteria for declaration programs. This includes the necessity for independent verification of data, as outlined in ISO 14025, Environmental labels, and declarations—Type III environmental declarations—Principles and procedures (Mazzi et al. 2017).

The ISO 14040 and ISO 14044 standards define Life Cycle Assessment (LCA) as a methodology for assessing the environmental impact of products, services, and processes. LCA, a stand-alone tool for identifying life cycle hotspots, primarily serves to compare products, processes, and activities (Fig. 9.6). It is recognized as a top tool for formulating environmental policies and is actively employed in creating policies for biofuels, energy, and wastewater treatment. Key sectors benefiting from this tool include plastics, detergents, personal care products, and automobiles. Additionally,

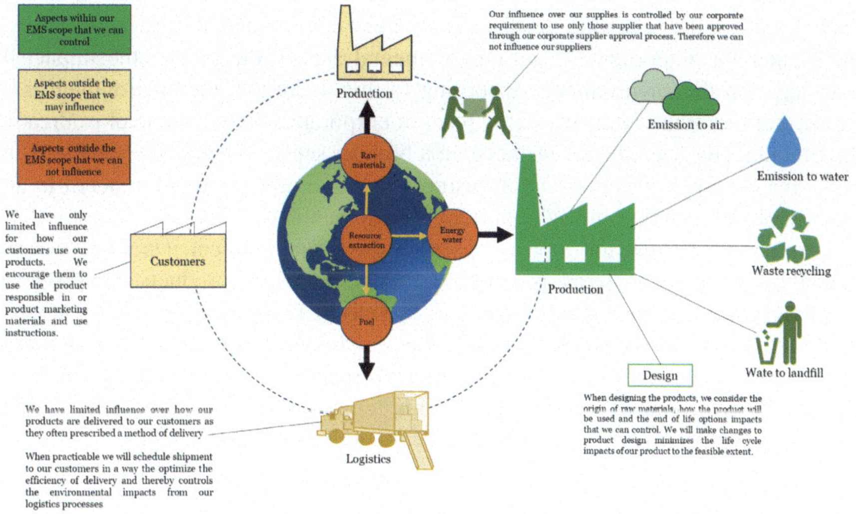

Fig. 9.6 Depicting lifecycle perspective

industries such as agribusiness, mining, oil and gas extraction, manufacturing, and retail commonly utilize LCA.

As well as improving the environmental profile of organizations, LCA and EMS are considered valuable tools; however, even though the EMS approach is procedural and the LCA approach is product-oriented, they can be complementary if we view them in terms of the comprehensive view. In the last few years, two different approaches have been proposed, both of which can benefit from both tools: integration and combination. When LCA and EMS are integrated, one tool's application field is expanded to include the main features of the other. Utilizing both tools coherently in a systematic manner in order to address the principal environmental management problems with an organizational-oriented perspective is the purpose of the combination. LCA, for example, can enhance environmental performance evaluation by adding objective and scientific elements and by providing insight into the burdens imposed by an organization within and outside its borders. According to some authors, the combination of the two tools offers the most promising solution. In other words, organizations that have already implemented an EMS or an LCA can apply the other tool and orient it to meet their specific needs and problems in order to complete the chosen tool.

9.5.5.3 Environmental Aspects in Product Standards

By the definition, "an environmental aspect is one that interacts or can interact with the environment as part of its activities, products, or services". Hence, the commentators on the definition point out that environmental aspects can only be defined as such if they have, or at the very least can have, at least one environmental impact. Generally, the greater the importance of an environmental aspect, the greater the impact it may have on the environment. According to this definition, the environment is a significant factor in practically every form of corporate activity. A parcel is already environmentally friendly, for instance, in a broader sense. When parcels are sent in bulk (for example, in a mail-order business), what might not be of concern in an individual case can become an important environmental concern.

In order to determine the impact of a service on the environment, it is not necessary to analyze every raw material, product, or activity. Activities, products, and services can be grouped or categorized meaningfully only if they share common features. Additionally, you need to determine whether any relevant aspects result from these characteristics, what significant environmental impacts may be resulting, and how to improve environmental performance. This "checklist" is not comprehensive and does not cover all aspects that apply to all companies and organizations.

- Environmental aspects as examples
- Atmospheric emissions
- Flows into water
- Soil contamination
- Raw material and natural resource consumption
- Energy consumption and release
- Waste generation
- Land consumption

A number of these aspects are at a higher level and may require further analysis. Whenever raw materials are consumed, their extraction and transportation must be completed first. In this case, each environment aspect is only partially able to be controlled but can be influenced to some extent. Indirect or direct environmental aspects are therefore determined by considering the product or service life cycle. In addition to the company's own products and services, third-party products and services must also be considered. In order to understand the terms activity, service, and product, let's first define them. In the core business, an activity is part of the production process (e.g., a step in the production process). An auxiliary service is defined as an activity that aids core operations, such as boilers, heating, cooling, and maintenance. The products you offer for sale are the goods you offer to the market. In terms of environmental impacts, a product's packaging, or its level of recyclability at the end of its lifecycle can be important environmental issues. According to ISO 14001, an organization must develop a process to pinpoint environmental aspects and recognize those that could have a substantial impact on or influence the environment. Additionally, you should keep a register of the environmental aspects, which should be kept up to date and reflect any planned changes or modifications to the activities, products, or services. Aspects can be classified as direct or indirect. In a direct environmental aspect, the organization has direct control over its own activities, products, and services (for example, waste management on site). For non-industrial organizations, however, indirect environmental impacts will tend to be more important (e.g., how subcontractors handle waste on your site, aspects controlled by chains, customer-controlled aspects). Your organization's activities, products, and services need to be studied for their environmental impact to identify your environmental aspects. Environmental aspects are typically considered together with other factors, such as pollution of air, water, and land, use of raw materials, waste, natural resources, and biodiversity impacts.

In identifying environmental aspects, not just the obvious core manufacturing or service activities, but all parts of the organization's operations must be considered. Each of these activities may have an impact on the environment, including maintenance departments, offices, canteens, heating and cooling systems, vehicle parking, contractor activities, and supplier activities. The environment at a facility can be assessed using a variety of methodologies—for example, value chain analysis, process flow analysis, identification of materials, and compliance with legal requirements. Ideally, key departments should be involved in a cross functional team. This will lead to an overall list of aspects and impacts for each type of activity, product, or service. Environmental aspects are evaluated by focusing on the most important factors. Managing all environmental aspects is not necessary—only the ones that are significant to you according to your own criteria. Environment management systems in your organization focus on important environmental aspects.

In order to determine the significance of environmental aspects, an organization can employ a variety of methods, depending on its type, nature, and complexity. You should consider the following factors when assessing significance:

- A potential threat to the environment
- The aspect's size and frequency
- Stakeholders' importance
- Legislation relevant to the environment

Through a systematic review of its environmental aspects and their current and potential effects, every organization can define its own significance criteria.

Summary

When it comes to environmental performance, organizations need to consistently aim for progress, one approach is to introduce is an EMS. Environmental management involves overseeing and controlling human effects on the environment, focusing on minimizing adverse impacts like pollution, habitat destruction, and climate change. An efficient EMS helps companies address environmental risks by offering a structured approach to managing environmental influences, outlining roles and responsibilities for all employees, and setting up procedures for testing and environmental inspections. This chapter detailed the concept of environmental management system, how it works, its benefits, importance and many other aspects. In recent years, there has been a growing emphasis on businesses operating in an environmentally responsible manner. Environmental management systems assist companies in recognizing and controlling their environmental effects, as well as monitoring advancements toward sustainability objectives. A successful EMS should comprehensively grasp all facets of its environmental performance. By utilizing an EMS, your business can reach its environmental goals and establish a cost-effective compliance plan. This system offers a proactive approach to achieving environmental objectives, reducing risks, and enhancing health and safety measures.

Questions/Exercise

1. What is ISO, ISO 14000, ISO 140001?
2. Enlist benefits of an Environmental Management System.
3. Describe the importance of EMS. Explain PDCA cycle of continuous improvement.
4. What are the objectives and organization standards in EMS.

References

Chromeextension://efaidnbmnnnibpcajpcglclefindmkaj. http://www.jiwaji.edu/pdf/ecourse/env_science/EMS%20Audit%20and%20mgmt%20review.pdf

efaidnbmnnnibpcajpcglclefindmkaj. https://www.iso.org/files/live/sites/isoorg/files/store/en/PUB100323.pdf

https://app.croneri.co.uk/topics/ems-standards/indepth

https://app.croneri.co.uk/topics/environmental-performance-evaluation-epe/indepth#:~:text=EPE%20is%20an%20ongoing%20process,tracking%20performance%20trends%20over%20time.

https://asq.org/quality-resources/environmental-management-system

https://asq.org/quality-resources/iso-14000

https://www.apec.org/docs/default-source/Publications/2003/2/Best-Practices-of-International-Organization-for-Standardization-ISO-14021-Self-Declared-Environment/03_cti_bestpracticeiso.pdf

https://www.epa.gov/ems#:~:text=An%20Environmental%20Management%20System%20(EMS,and%20increase%20its%20operating%20efficiency

https://www.gdrc.org/uem/iso14001/info-13.html

https://www.process.st/iso-14000/

Jasch C (2000) Environmental performance evaluation and indicators. J Clean Prod 8(1):79–88

Mazzi A, Toniolo S, Catto S, De Lorenzi V, Scipioni A (2017) The combination of an environmental management system and life cycle assessment at the territorial level. Environ Impact Assess Rev 63:59–71

Rocha C, Silvester S (2001) Product-Oriented Environmental Management Systems (POEMS) from theory to practice–experiences in Europe. Project "Start IPP" e Starting with the Integrated Product Policy

Chapter 10
Environment Risk Assessment (ERA)

10.1 Introduction

Environmental risk assessment (ERA) is a systematic approach to identify, evaluate, mitigate, and control hazards that may arise as an outcome of process, procedure, or equipment failures. Demanding industrial processes lead to an increase in industrial accidents, public scrutiny, loss of life and property and legislative obligations. Hence it became necessary to ensure that risk management is undertaken and implemented effectively. ERA is not only a necessary outcome of environmental scientific advancement, but it is also a critical requirement for societal security. It has evolved into a scientific and vital foundation for environmental risk management and decision-making. An appraisal of what the risk means in life of the affected person may be included in the risk assessment. This is greatly dependent on how the risk is perceived (Aven 2015). People's ideas, attitudes, judgements, and sentiments about risks are all part of risk perception, as are the broader cultural or societal values that public hold about hazard and its benefits. People's perception about risk is critical in assessing and managing process of risk. Perception of risk will play a significant role in determining if a risk is deemed "acceptable" and if risk management techniques are seen to be successful in resolving the problem.

In different kind of professions and academic fields risk assessment technique is utilised. Whilst social welfare professionals "risk assess" their clients to determine the likelihood of anti-social behavior repeating, engineers "risk assess" bridges in order to identify the possibility and consequences of component failure. Risk assessment is performed in order to make a risk management choice. It has become a popular method for investigating environmental issues. It is used to investigate dangers of various types. It has been claimed that the process of assessing scientific risk should be kept apart from the process of managing risks for policies, although this is no longer universally accepted (Dubois 2010). They are inextricably related. Risk management is a judgement making process that allows selections to be made among a variety of

© The Author(s), under exclusive license to Springer Nature Switzerland AG 2024

R. Bhateria et al., *Environmental Impact Assessment*,
https://doi.org/10.1007/978-3-031-66797-8_10

choices that meet the "required outcome." The "required outcome" could be established via formalized risk-cost–benefit analysis, industry norms or "good practice", or it could be established by law in the form of enforceable environmental regulations. The risk management approach defines risk as the degree to which a hazard has negative impacts while accounting for the likelihood of potential harm and side effects:

$$\text{Risk} = \text{Probability} \times \text{Consequences}, \text{ or } R = Q \times P$$

where,

$$R = \text{Risk},\ Q = \text{probability (probability of an accident)},$$
$$P = \text{consequences (amount of the loss)}.$$

This formula serves as the foundation for variety of quantitative risk analysis approaches, and it can be used for any unfavourable occurrence. Therefore, it is possible to evaluate the operation's overall risk by compiling the risks under all conditions. In risk management, the unit of recovery is equal to the measurement unit, and the risk is the mathematical potential of repercussions.

This equation can also be used to characterize risk:

$$''R = E \times A \times S''$$

This Equation consists of:

$$E = \text{probability that there are certain risks},$$
$$A = \text{probability of avoiding risks},$$
$$S = \text{Category, which determines the severity of risk}.$$

Both quantitative and qualitative methods of estimating risk take into account the findings of measurement and computation. The amount of harm effects and risk of effects are identified in the suitable measurement unit using quantitative risk assessment. While qualitative risk assessments identify the origin of potential hazards, quantitative estimates of risks are based upon mathematical calculations. The amount of loss or repercussions, as well as the likelihood that the event will be dangerous, are not actually determined by qualitative risk assessment. Building a set of metrics and a framework for risk management, as well as determining the degree of hazard objectively, are made feasible by numerical risk assessment (Eduljee 2000).

10.2 ERA and EIA

Both ERA and EIA are structured tools that provide decision-makers with recommendations about the environment for the implementation of projects that may be extended to strategic levels of decision-making. Although they both are similar but have some fundamental differences. As ERA does not include the consideration of development alternatives but EIA does. Both approaches are iterative and when a project or proposal is implemented in final stages both monitoring and auditing are required. While ERA promotes risk perception and risk communication, EIA places more emphasis on public participation and consultation. While ERA considers the occurrence and extent of adverse effects, EIA focuses on the initial query regarding possible consequences for the environment. While environmental impact assessment encompasses a broad range of activities, risk assessment focuses on specific regulatory issues and employs quantitative analysis to assess the probability of undesired events. By measuring uncertainties, evaluating the likelihood that an effect will occur or that environmental criteria will be exceeded, and identifying the need for abatement or additional research, risk analysis, especially when handling uncertainty, greatly aids in environmental assessments.

The systematic process of assessing ecological impacts that may arise or are now occurring as a result of organisms and populations being exposed to one or more chemical substances is known as Environmental risk assessment (EPA 2016). To assist in informing planning and decision-making processes, the ERA approach entails the identification, analysis, and presentation of information in terms of risk associated with environmental values. It is not intended to replace the role of planning and management processes and does not promise to provide all economic and social information necessary for making decisions. The framework for ERA and EIA includes the salient features that are tabulated in Table 10.1.

10.2.1 Similarities Between ERA and EIA

The fundamental EIA procedure can be integrated with the framework of environmental risk assessment (ERA). The basic EIA process and the ERA framework coincides in principals, making integration at all policy and regulatory levels possible. For example, both frameworks (Zeleňáková and Zvijáková 2017; Antunes and Santos 1999).

- "aim to strike a balance between environmental quality and ecological functions in order to promote sustainable development."
- "assist in the development, implementation, and evaluation of policies that support sustainable development,"
- "can be applied to different analytical scales, such as local, regional, continental, and global scales," and

Table 10.1 Framework for ERA AND EIA

Framework for ERA	Framework for EIA
Screening: Screening is the first step that is done to determine risk range and identify the various factors that may pose significant damage to the environment. It is cost effective and efficient procedure that helps to prevent unnecessary and resource-intensive assessments for low-risk activities or substances while concentrating resources on environmental issues	**Screening**: For any project or proposal preliminary assessment of the existing data related to environment is done under screening to decide whether detailed EIA is required followed by scoping, while smaller projects are exempted from further assessment
Hazard analysis: It involves the systematic identification and assessment of potential hazards or adverse effects that may arise from exposure to particular stressors or substances. It lays the groundwork for the exposure assessment, dose–response assessment, and risk characterization stages. It offers essential information that helps decision-makers to evaluate the overall risk and identify the most appropriate risk management strategies	**Baseline studies**: Baseline studies assess the potential environmental impact of a proposed project. It involves collection and analysis of data on existing environmental conditions including physical, socio-economic and biological factors in and around the area before starting the project
Risk evaluation: It includes systematic analysis and quantification of the possible risks identified during hazard identification, exposure assessment and dose–response assessment. It is done to assess the probability and scale of adverse effects on environment or human health and to inform the decision makers about strategies for risk management	**Assessment and Evaluation**: It is systematic and comprehensive examination of the possible environmental, social, and economic impacts related to the project or activity under consideration. It involves collecting data, conducting studies, and evaluating information to understand the possible effects of the project on the environment and human well-being
Monitoring and Audit: After completing all the steps monitoring was done to ensure environmental risks related to any project and helps in preventing and mitigating environmental damage and further leads to confirmation and rejection of the project	**Monitoring and Audit**: After evaluation all the factors are considered and based on that the project is confirmed or rejected

- "are adaptive and considers problem assessment, policy priorities, formulation and implementation of policies through adequate tools, and takes into account the perspectives of the stakeholders involved."

EIA procedures and ERA both involves consultation with stakeholder and establishment of context. The risk-based EIA involves integrating two procedures in order to benefit from risk analysis while satisfying legal standards for statutory assessment procedures and promoting the application of modern environmental protection techniques.

10.3 Environment Risk Assessment

Our main worry is the likelihood that exposing a certain population to a certain combination of chemicals would result in a certain reflex, such as malignant tumor, infertility, neurological impairment, developmental issues, or birth deformaties. Hazard identification, dose–response assessment, exposure assessment, and risk characterization are the four stages that the National Academy of Sciences recommends be included in the risk assessment process. The crucial stage of risk management comes next once a risk assessment is finished, as depicted in Fig. 10.1.

10.3.1 *Hazard Identification*

Finding out whether a certain chemical is directly related to specific health outcomes, such as cancer or birth defects, is the process of hazard identification. This step mainly concentrates on whether a substance is hazardous to animal or other experimental organisms because human data are frequently difficult to get. The "Red Book" defines hazard identification as the process of determining if exposure to a chemical can raise the likelihood of a negative health consequence. Depending on the nature of the hazard, many methods are utilised to assess risks, including observational epidemiologic research, animal and perhaps human laboratory tests, mechanistic investigations, and other mechanistic experiences (Dinis and Fiúza 2009). Epidemiological data has certain inherent flaws as a result of its observational character. For instance, only when the health effects of contaminants are actually experienced can the threat of contaminants be evaluated (Swaen 2006). Furthermore, differences among different factors that are connected to the risk could throw the results off.

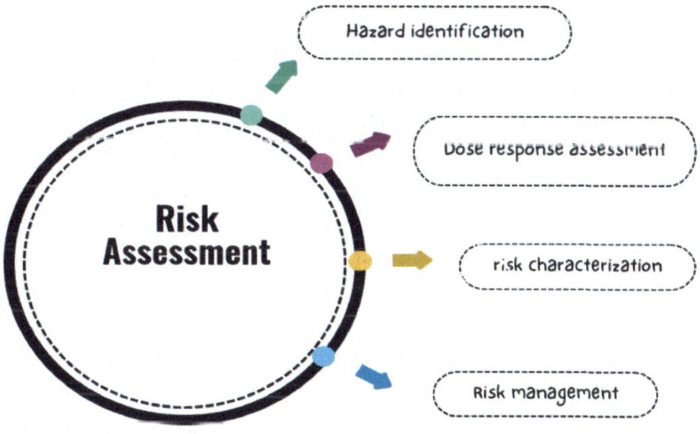

Fig. 10.1 Stages involved in environment risk assessment

Confounded behaviour has a negative impact on epidemiological research and can lead to inaccurate findings.

Additionally, epidemiological research is costly and frequently necessitates a lengthy time frame to observe its risks (Nachman et al. 2011). Epidemiological study has the ability to examine the effects of contaminants on diverse populations, evaluate cumulative hazards and better reflect temporal mechanisms. It can also address exposure pathways that are pertinent to the general population and is more appropriate for taking into account the actual range of population exposures, despite its inherent limitations. Experimental human studies can be affected by people's lifestyles, places of employment, diets, social affiliations, and a variety of other factors that could affect people's health. Furthermore, the scope of such investigations is restricted to brief exposures to pollutants that are not anticipated to have long-term negative effects (Nachman et al. 2011). Additionally, it can be challenging to get volunteers for the study, for instance, in securing ethical approval (Artal and Rubenfeld 2017). As a result, experimental investigations of humans are rarely carried out. The application of animals used in laboratory, particularly mice and rats, to ascertain chemical dangers is a crucial part of toxicology and chemical risk assessments because experimental human investigations are rarely conducted. The main benefit of using animals in studies as opposed to people in research on humans is that their exposure can be empirically tracked. The utilization of laboratory data from animal to predict human health hazards is, however, fraught with several sources of ambiguity, including questions about the applicability of animal model findings to humans. Identification of the hazards defines the type and strength of the evidence, which requires considering causality. Evidence for hazard identification may come from epidemiological or other human research, outcomes from laboratory animal testing, in vitro analyses of cells with both human and animal origins, quantitative structure–activity estimations, or other in silico models. By itself, hazard identification can lead to the conclusion that a substance poses little to no risk to human health or the ecosystem and shouldn't be subject to regulatory oversight.

The hazard identification process normally enables the determination of the concentration with a specified environmental effect that is important but assessed under experimental settings (no observable effect level, 50% lethal concentration (LC50), or 50% effective concentration (EC50) (Pellizzato 2014). Mechanistic investigations describe how particular negative effects are brought about by a chemical and are conducted in vitro and in vivo in laboratories using cellular, biochemical, and molecular processes as their basis. It is possible to determine which cellular, metabolic, and molecular processes in test organisms and humans are comparable or different, which supports the credibility of extrapolation. However, distinct systems may contribute to a certain end point for a given chemical, and it may not always be clear how different approaches interact with different organisms to yield unfavorable outcomes (Zhang et al. 2023).

Hazard Identification and Risk Analysis (HIRA) is a comprehensive term that encompasses all operations related to identifying hazards and assessing risks throughout a facility's life cycle. Its goal is to protect workers, the general public, and the environment by ensuring continuous risk control within the parameters of the organization's risk tolerance. Identifying hazards, evaluating potential repercussions, and estimating the chance of their occurrence are the three primary risk questions that HIRA normally answers. The goals, information, and resources at hand determine the level of analysis needed. Depending on the level of complexity required, a variety of methods and approaches are available for hazard identification and either quantitative or qualitative risk assessments. Risk analysis and hazard identification can be done using a variety of instruments, from fundamental qualitative methods like checklist analysis and hazard and operability analysis (HAZOP) to more intricate quantitative methods. The objectives, information, and resources at hand determine the precise amount of analysis that is needed.

The remaining three elements of the risk assessment process includes exposure assessment, dose–response assessment, and risk characterization, which need to be carried out if a chemical was found to be potentially harmful. It's crucial to distinguish between hazard identification and hazard characterization; hazard characterization might be a better name for dose–response assessment. Traditional methods frequently require a large scale of animal experiments that can consume a variety of resources in terms of evaluation and research (Galimberti et al. 2020). For the majority of organic contaminants, there is little or no toxicology data available (Chi et al. 2018). The initial step in a risk study is to determine to what extent the substances to which an individual has been exposed have a probability of having any detrimental effects on health. Toxicologists investigate the types of harmful effects that poisonous chemicals can induce as well as the possibility that they occur. Before detailing this danger identification process, we first briefly review the many routes a chemical may take via a human body and the types of health effects that may occur. A toxin can enter the body in three different ways: by ingesting food or liquids, by inhaling, or by coming into touch with the outermost layer of skin or other external surfaces, like the eyes. After intake, it may be transferred to numerous organs and systems by the blood after being absorbed by the body. The toxicant may subsequently be cleared from the body by excretion or by changing into another substance, or it may be kept (for instance, in fat, as in the case of DDT). Figure 10.2 shows fate of toxins in the body. Although metabolism can change chemicals into more hazardous forms, the biotransformation process typically produces metabolites which are quicker to leave the body than the original molecules.

Toxins can particularly affect a few organs. For instance, the liver is frequently the target since it filters blood before it is pushed through the lungs. Toxins can directly harm the liver because they are delivered by the bloodstream and because it receives so much blood supply. Furthermore, because the liver's primary job is to metabolise substances and change them into forms that the body can eliminate more quickly, harmful compounds created during the biotransformation process themselves can also assault the liver. Hepatotoxins are substances that can harm the liver. Hepatotoxic compounds include several manmade chemical molecules such as chloroform, trichloroethylene; carbon tetrachloride; pesticides; drugs and heavy metals can damage kidneys. It is incredibly thrilling to study the science of risk analysis in the early 21st century. It is clear that humanity needs this science to be applied to environmental issues. The increasing modernization of toxicity testing resulted

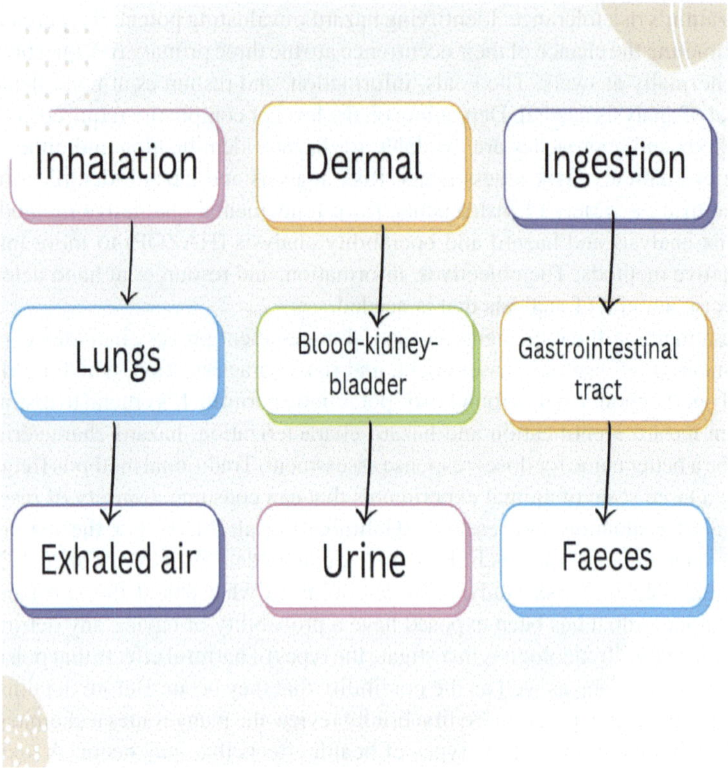

Fig. 10.2 Fate of toxins inside the human body

from the realisation that approaches and procedures frequently utilised in the pharmaceutical sector could transform toxicology. But then came the difficulty of putting these strategies into practice. There is still a lack of confidence in these approaches' forecasting abilities. Therefore, it is unknown how much these non-animal testing techniques are taken into account during hazard identification, problem formulation, planning and scoping.

10.3.2 Hazard Assessment

Hazard assessment include peering over the work area, assessing tools and procedures, and spotting possible risks. Hazard assessments can be done in two ways: site-specific evaluations that concentrate on a specific worksite during a specified period of time, and formal assessments that encompass every task inside an organization. Hazard assessments are used to create plans that evaluate risks, identify potential hazards, and set up suitable controls depending on particular working scenarios.

The onus of spotting possible health and safety risks falls on both employers and employees. Before every shift or activity, a site-specific hazard assessment is carried out in industries with dynamic activities and conditions. To protect employee's health and safety, danger assessments must be carried out at least once a year by stable firms. This is just as critical as the danger identification step in the monitoring and evaluation process.

10.3.3 Dose Response Assessment

The process of evaluating the relationship between the dose of an agent given or received and the probability of a negative health consequence is known as dose response evaluation. A toxicological threshold for pollutants is produced in dose–response analysis and represents an estimate of the sustained exposure to a population that is doubtful to result in significant adverse effects over the course of its lifetime. A wide number of dose–response relationships are available for any particular medicine, depending on aspects such as whether the reaction is carcinogenic or not, and whether the test is a one-time acute test or a long-lasting chronic test. Due to the fact that the majority of studies are carried out at high doses, the right way of data extrapolation to lower exposure levels that people are most likely to encounter must be taken into account throughout the dose–response evaluation. The evaluation must also contain a strategy for extrapolating animal findings to people. It is also typical for toxic chemicals to damage other organ systems and organs. The kidneys' function is to filter blood and separate waste products that will be eliminated through urine. Nephrotoxins, or toxicants that damage the kidneys, include particular amounts of chlorinated hydrocarbons and metals including lead, mercury, and cadmium. Extreme harm to the kidneys can restrict or halt urine flow, which can lead to body poisoning from waste materials produced by the body and death. The term "hematotoxicity" refers to a substance's hazardous effects on the blood. Some hematotoxins, such as nitrates in groundwater and carbon monoxide in contaminated air, interfere with the blood's capacity to transport oxygen to tissues. Other toxic substances, like benzene, interfere with platelet development, which is essential for blood clotting. Due to their close proximity to pollution, the skin and lungs are frequently harmed by chemical toxins. Smoke from cigarettes, ozone, asbestos, and quartz rock dust are just a few examples of the pollutants that might damage lung function. Chemical toxins can cause the skin to respond in a number of different ways, but the most prevalent and dangerous skin issue related to the environment is cancer brought on by too much UV light.

It is common practice to presume that exposure to any amount of a drug that causes a carcinogenic response will increase the risk that someone will develop cancer. In order to navigate through the origin, a response versus dose plot is necessary. It is typically believed that there is a threshold dose below which there will be no noncarcinogenic reactions. The dose–response curves as well as the applied to them are techniques consistent with these two presumptions. There are different effects that cause cancer and those that do not. By the way, it's possible that the same molecule can trigger both types of reactions. Human dose–response data from animal bioassays must be scaled up in order to be applied. Assuming that the amount given per unit of body mass index in both humans and life forms is the same, the scaling factor can be constructed on the premise that dosages are equal. When scaling up from an animal to a human, identical doses are occasionally normalised to body surface area instead of body weight, especially if the exposure is dermal. In either scenario, the conventional dose units of mg/kg/day are used to specify the resulting human dose–response curve. To take into account variations in chemical absorption rates, adjustments may also need to be made between the responses of animals and humans.

10.3.3.1 Reference Dose

Imagine there is a precise threshold for a certain toxin in a specific animal species. We might envision a testing programme where animals would be exposed to various doses in order to experimentally discover the threshold. A dose below the threshold would not produce a response, whereas a dose beyond the threshold would. Although it is commonly believed that carcinogenic pollutants have no threshold, other contaminants do (Swartjes 2015). The US EPA calculates the exposure dosage (also known as the reference dose, or RfD) for pollutants with a threshold or concentration (reference concentration—RfC) as the level below which there will be no health concerns for people. The United States Environmental Protection Agency (US EPA 2022b) defines RfD is a forecast of a daily oral dose to the human population (including sensitive subgroups) that is probably not going to have a major lifetime risk of adverse effects. RfC is described as an estimate of a lifetime continuous inhalation exposure to humans (which includes sensitive subgroups) that is anticipated to be free of a significant risk of deleterious effects (U.S. EPA 2022b).

According to Barnes and Dourson, the RfD and RfC are determined by dividing the lowest observed adverse effect level (LOAEL) or no-observed adverse effect level (NOAEL) from animal experiments (or human studies, if available) by a set of uncertainty factors based on a technical assessment of the data's adequacy. Lowest Observed Effect Level (LOEL) refers to the lowest dose supplied that elicits a reaction. The no-observed-effect level (NOEL) is the greatest dose administered that does not elicit a reaction. NOELs and LOELs are frequently improved by distinguishing between impacts that are damaging to wellness as well as effects that are not. The acceptable daily intake (ADI), formerly known as the RfD is meant to provide an indication of the amount of human exposure that is predicted to be free from significant danger. RfD is measured in the same units as chronic daily intake (CDI): averaged over a lifetime, mg/kg-day. Divide the NOAEL by an appropriate uncertainty factor, commonly referred to as a safety factor, to get the RfD. For pollutants lacking a threshold, the unit risk value is used to determine the risk of carcinogenesis. Although it has been used for many years in dose–response assessment, the NOAEL/LOAEL technique has drawbacks. The primary issue with the current toxicity reference value database is the inconsistent format of the data content, the absence of complete data, and the uniformity of the toxicity database across various populations (Zhang et al. 2023).

10.3.3.2 Exposure Assessment

Exposure assessment describes the frequency, duration, and quantity of substance exposure that an organism may experience. It is based on data from modelling and monitoring as well as data on environmental fate and mobility. The size, composition, period of exposure, and toxicant concentration at which the population was exposed are all important factors to consider when estimating exposure to a given toxicant. It is crucial to take into account the population's age and health, smoking history, possibility of pregnancy, and whether or not multiple exposures to toxins could have additive effects. An exposure pathway is the route that a potentially hazardous material takes, via environmental carriers or media, most commonly air (volatile chemicals or particles) or liquid (soluble or colloidal compounds), from its source to a receptor (such as a person or an animal). Electromagnetic radiation is an exception because it does not require any medium. The exposure route, often referred to as the intake pathway, is the mechanism by which the transfer occurs; it usually entails ingesting, inhalation and or skin contact (Fig. 10.3). Direct contact can affect a system as a whole or just locally at the point of contact.

The accumulating effects of prior toxic exposure, particularly interaction with soil and groundwater, is referred to as the contamination load for human health at a

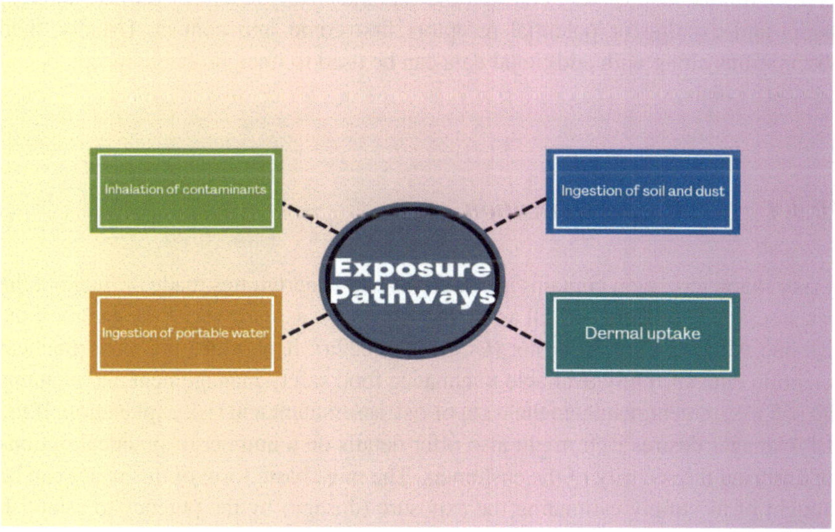

Fig. 10.3 Potential routes of pathogen exposure

contaminated site (Swartjes 2015). Four main concepts, starting with the most basic theory, which control the movement and fate of pollutants in soil and groundwater are advection, dispersion, transition mass transfer, and transformation processes (Brusseau and Chorover 2019). A human exposure evaluation is a two-step process. Prior to coming into contact with people, harmful chemicals must first travel along specific paths that must be assessed. Second, the amount of contact that people are expected to have with those contaminants must be estimated. When compounds are exposed to the atmosphere, plume models are frequently used because they may volatilize and move with the wind in such circumstances. When something comes into contact with soil, it may seep into the groundwater, where it may eventually find its way to nearby drinking water wells. Pollutants may undergo numerous changes during transportation that may alter the toxicity and or concentration. An estimation of the toxicant concentrations in the air, water, soil, and food at a specific exposure point can be made once the exposure pathways have been examined. The second phase of an exposure evaluation starts once the concentration of various hazardous substances have been determined. There must be an estimate of human contact with pollutants. The number of persons exposed, how long they were exposed, and how much soil, water, food, and air were polluted that enter each exposed person's body are all necessary pieces of information. An equation containing three sets of variables can be used to quantify exposure, intake, or potential dose:

- Media containing high levels of chemicals or microorganisms
- Exposure rates (intensity, frequency, and length)
- Quantified biological properties of receptors, such as body mass, chemical absorption capacity, and degree of tolerance to microbial diseases

Exposure concentration is calculated using data that has been measured, monitored, or modelled. Exposure concentration should ideally be assessed where ambient media and existing or potential receptors first come into contact. Usually, field observations along with additional data can be used to trace probable receptors and exposure points.

10.3.4 Risk Characterization

A risk characterization contains assumptions and uncertainties made throughout the risk assessment process as well as the risk assessor's assessment of the nature of the risk and whether it exists or not (U.S. EPA 2022a). It provides the most practical scientific data currently available to enhance food safety management and contains the risk assessment results in the form of risk assessment and risk explanation. If the risk manager desires it, it might also offer details on a number of practical options for ensuring the security of the customers. The most basic form of this phase can be thought of as simply estimating the exposure (dosage) by the potency to establish an individual's risk and multiplying that result by the total number of exposures to produce an estimate of the population's overall danger. The positive outcome and expenses of risk mitigation are the factors to be considered while comparing the risks. In cases, where the details provided by the impact assessment have not sufficiently addressed the size of the hazard, the risk characterization may only serve as a preliminary evaluation. When the impact assessment depends on a threshold test rather than a full dose/response assay, this element is very important.

10.3.4.1 Uncertainty Assessment

Every stage of the risk assessment process contains some element of uncertainty. Therefore, we need to have a general understanding of the type and degree of uncertainty in the risk assessment before we can start to characterise risk. Different types of uncertainty factors: UFinter, the first type, takes into account the distinctions between the tested species and humans. In comparison to other examined species, humans may exhibit different levels of sensitivity to a given drug. The enhanced sensitivity of some human subpopulations to a chemical in comparison to the general human population is addressed by the second uncertainty factor (UFintra). When there is insufficient data or scientific understanding, a "no effect" threshold is extrapolated from a LOAEL, giving rise to the third type of uncertainty (UFother). The safe level (RfD) becomes closer to zero, more are the uncertainties because these UFs are in the denominator of the computation. Uncertainty factors are usually multiples of 10, while UFother can be anything between 2 and 10.

Errors in interpretation can also give rise to uncertainty and result in false positives and false negatives. When an issue is real but is mistakenly reported as not existing, this is known as a false negative. Environmental and public health organizations, along with advocacy groups, frequently tackle this problem by posing queries regarding various scenarios, including the biosensors' detection limit, possible leaks even when detector readings are zero, the accuracy of carcinogenicity tests, exposure via various routes, and the intricacy of mixtures. However, false positives present a problem for public health organizations who are in charge of shielding the public from environmental toxins. This happens when a substance is incorrectly classified as a possible risk or hazard due to flawed models or incomplete knowledge, creating unwarranted worries and diverting resources. False positives have the potential to damage decision makers' reputation and professional trust, which can affect public confidence.

Other sources of uncertainty are extrapolation from high to low doses; extrapolation from animal to human responses; conversion from one exposure pathway to another; limitations of analytical methods to quantify the organism; and exposure estimations. Even if they are often substantially bigger in assessments of exposure and the correlations between dosage and response (for example, % mortality), the uncertainties resulting from all the stages in a risk assessment should be considered into account as part of risk characterization (Gerba 2015). Due to the current lack of understanding regarding the behaviour of contaminants, uncertainty is unavoidable (Dong et al. 2015). Risk assessors are extremely worried by the lack of clarity surrounding the parameters utilized in the risk assessment process, the simplification of the assessment approach, and the improper selection of assessment scenarios (Chen and Ma 2006).

The outcomes of the risk assessment and the ultimate risk management choice depends greatly on having a thorough understanding of the uncertainty in the process. This is because uncertainty highlights the drawbacks of the assessment conclusion and explains the range of possibilities for implementing the risk assessment results. The entire risk assessment process which includes hazard identification, exposure assessment, dose–response assessment, and risk characterization is filled with uncertainty (Dong et al. 2015). Sensitivity assessments are often used to describe uncertainty. In sensitivity analyses, each parameter's uncertain quantities changes with respect to average values, high and low estimates (one at a time) to observe how they impact the final risk assessment. This procedure outlines those variables which are most important in finding out the extent of the risk along with the range of probable values for the risk as a whole. As long as the exposure remains below the health-based reference value, it is widely acknowledged that a chemical does not pose harm to human health. A choice can be made regarding any required management action, if any. The risk manager may occasionally request more information from the risk assessor on the risk assessment. For instance, the risk manager may ask for the expected health effects of various maximum authorised levels of a contaminant found in food to enable balancing the danger of a chemical with the health advantages of food accessibility (Moy 2014). A good risk assessment reiterates the assessment's objectives, effectively communicates the findings, and distinguishes between scientific and political conclusion. In essence, it explains why all of the decisions taken during the risk evaluation procedure led to the outcome of the risk assessment (U.S. EPA 2000).

10.3.5 Risk Management

The public's response to a risk that has been identified, evaluated, and categorised depends on how they interpret and comprehend the data. Determining the possibilities that exist, risks, costs, advantages, and their distribution throughout the society should be stated in the decision analysis, which should be the framework through which risk information is delivered. In order to set priorities, risk management requires assessing the cost and effectiveness of risk reduction methods. Risk management involves assessing the available risk-reduction strategies and putting the most cost-effective ones into practice (Galatchi 2006). Three important phases of risk management are shown in the Fig. 10.4.

It is helpful to examine two well-established pillars of risk management before assessing recent advancements in core concepts and techniques: (a) the main risk management approaches that are now in use, and (b) the structure of the risk management process. The three primary tactics usually used to manage risk for the main risk

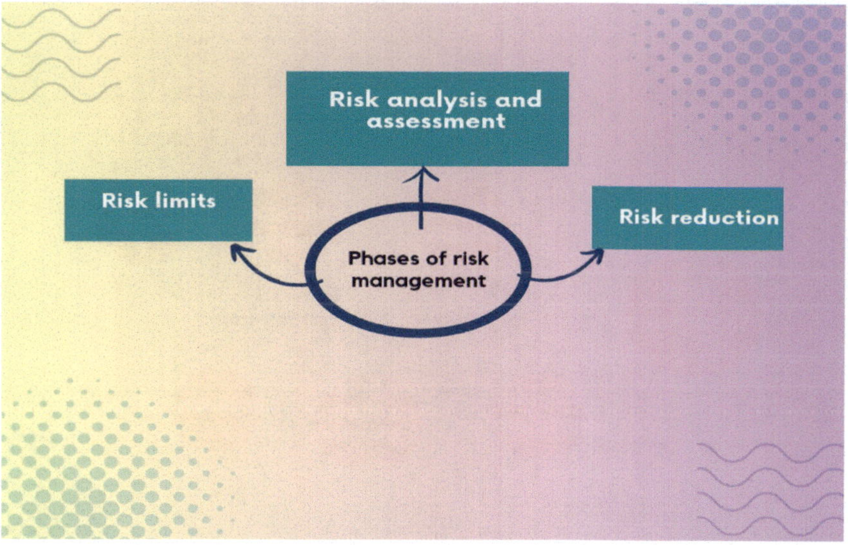

Fig. 10.4 Three different phases of risk management

management approaches are hazard-informed, precautionary, and discursive techniques (Renn 2008). The cautious approach is another term for a strong and resilient strategy. A mix of these three strategies is usually the best course of action. When there are significant risks to the well-being of humans or the environment and there is conflicting scientific evidence, decision-makers can use the precautionary principle to guide their actions.

In this situation, being able to accurately identify the signs and symptoms of important occurrences is essential. Every risk law is, to some degree, predicated on these ideas in order to address the uncertainties, dangers, and potential for surprises. The discursive technique tries to improve trust and confidence by removing doubts and inconsistencies, clarifying facts, engaging those who will be affected, engaging in dialogue, and holding people accountable (Renn 2008). "Risk-informed strategy" refers to risk management, which uses either absolute or comparative risk assessments to determine prevention, removal, transfer, and maintenance. An overview of today's situation of food safety is usually found in a risk profile, which also includes information about the risk itself, how and where it enters the food supply, how frequently, where it is found, and how much of it is present in food, as well as information about customer's opinion of the risk and ways to minimize it (Moy 2014). Risk assessments are helpful in decision-making when choosing between possibilities, accepting things and activities, implementing risk-reduction measures, etc. Producing risk information frequently includes the addition of decision analysis methods including cost–benefit evaluation, multi-attribute evaluation and efficiency assessment. While all of these methods are systematic ways to organise a decision

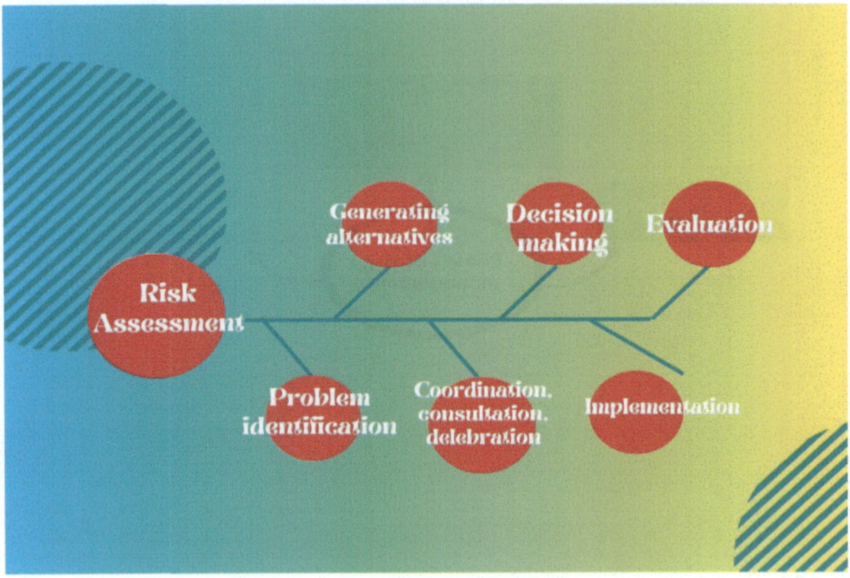

Fig. 10.5 Stages of developing a policy for risk management

option's benefits and drawbacks, they differ in how precisely similar they want the problem's factors to be.

Policy analysis and risk management are inextricably linked. The concept of "policy" applies to both world organisations and governments, as well as private sector companies and institutions. It is an idea or plan that guides decisions and achieves desired outcomes. The phases depicted in Fig. 10.5 are frequently used to coordinate the development and execution of policies. The risk field contributes to the many steps in the policymaking process, firstly by defining and describing the problem or issue, taking into account factors like goals, standards, risks, uncertainties, and knowledge, and secondly by outlining and emphasising fundamental concepts (such as the precautionary principle) and how to combine development and growth (Aven 2016). Risk assessment and management science is well-established, and it contributes significantly to practical decision-making.

10.4 Advantages of ERA

Over the past 20 years, ERA acts as an effective set of tools for analysing negative impacts on environment, has also been used in aiding the decision-making process in EIA. Risk management is the action done after taking into account this and other information, while risk assessment gives information on potential health or ecological concerns. Techniques for identifying and evaluating risks can be simply adapted to incorporate negative environmental consequences. Within an EIA, ERA has two main benefits: it provides a mechanism for addressing uncertainty and establish EIA priorities; it also identifies potential hazards associated with a proposal in a methodical manner and enhances the area of the comprehensive investigations needed for the EIA.

The major use of ERA in EIA is to help during investigation scope. The most significant project threats are frequently found during the EIA process' scoping phase, which additionally occurs when the study's general design is developed. The scope and complexity of possible EIA studies are specified during the scoping phase. An effective approach for determining which hazards should be investigated in-depth and which should be isolated from other issues that simply need to be given little attention. Some advantages of implementing ERA during the scoping stage include the following:

- Increases confidence that all relevant risks requiring investigation have been recognised;
- Informs the development of the planned research programme;
- ERA makes sure that the quantity of inquiries equates with the risk; and
- ERA increases the EIA process's efficiency.

Two more important uses of ERA involve dealing with uncertainty and setting priorities for EIAs. During the EIA approach, ERA can play a crucial role in minimising uncertainty. The ERA technique, which takes into account probable worst-case situations, can be used to explore uncertainty as part of the process. The risk assessment process integrates uncertainty by allocating probability and impact levels. This way, the procedure embraces uncertainties by establishing probability and impact levels. If it is unknown whether the vegetation to be destroyed provides adequate shelter for an endangered species, the risk analysis should assume that it does. Although the technique can exaggerate threats, it does provide a maximum threshold for risk assessment and produces conservative risk levels. This cautious strategy is beneficial since it can demonstrate that risk is not significant (even if uncertainty is considered). Alternately, it can show that a risk is possibly large based on a cautious evaluation. This could lead to: additional data collection or inquiry to lessen uncertainty; research into alternate project execution strategies that more effectively handle the problem; controlling uncertainty during the project conceptualization and implementation phases (Elliot and Thomas 2009).

Limitations of ERA in EIA

Understanding the limitations of ERA techniques determines how the results should be used in the EIA and improves comprehension of the level of accuracy of the risk assessment output. The risk management process includes setting the conditions and obtaining preliminary data, assessing the project and identifying risk routes based on conceptual frameworks, developing a risk matrix with description of the probability and consequence, assigning risk scores with expert input, often in the context of a risk workshop, reviewing and fine-tuning risk ratings, and identifying mitigation strategies to address major hazards (Burgman 2005).

Efficient pathways for cross-disciplinary collaboration and the possibilities for outside assistance and evaluation of risk evaluation data might decrease the downsides of ERA, where ERA is not appropriate, used in an EIA process. Additionally, ERA approaches are not applicable to a variety of impact assessment objectives. Accepting these ERA limitations ensure that it is used appropriately in the EIA. For instance, ERA is unable to make an impartial determination of whether a risk is acceptable. Because these evaluations are based on standards, there will be discussion on whether intermediary risks are acceptable. The majority of respondents will agree on risks that are manifestly unacceptable and hazards that are clearly undesired. The degree of risk aversion varies between parties. Moreover, ERA is not a good tool for assessing a project's compliance with regulations and policies as a component of the EIA process. This is because violation with policy or law may still happen, even if a problem is low risk.

Summary

A risk assessment often is an continuous process. There are mainly three steps which makes up risk assessment: identification, estimation, and evaluation. However, the risk management framework further has a fourth step, which requires feedback on each and every risk assessment project. Nonetheless, the current state of environmental hazard reduction is seriously lacking due to the absence of such input. This chapter's goal is to discuss hazards in relation to the risk management framework.

Questions

1. What is an Environmental Risk Assessment?
2. Why Environmental Risk Assessment is Important?
3. Discuss about the various steps in performing EIA.
4. How do Environmental decisions use Risk Assessment?
5. Write down some Environment Risk Assessment examples.

References

Antunes P, Santos R (1999) Integrated environmental management of the oceans. Ecol Econ 31(2):215–226

Artal R, Rubenfeld S (2017) Ethical issues in research. Best Pract Res Clin Obstet Gynaecol 43:107–114

Aven T (2015) Implications of black swans to the foundations and practice of risk assessment and management. Reliab Eng Syst Saf 134:83–91

Aven T (2016) Risk assessment and risk management: review of recent advances on their foundation. Eur J Oper Res 253(1):1–13

Brusseau ML, Chorover J (2019) Chemical processes affecting contaminant transport and fate. Environmental and pollution science. Academic Press, pp 113–130

Burgman M (2005) Risks and decisions for conservation and environmental management. Cambridge University Press

Chen YC, Ma HW (2006) Model comparison for risk assessment: a case study of contaminated groundwater. Chemosphere 63(5):751–761

Chi Y, Zhang H, Huang Q, Lin Y, Ye G, Zhu H, Dong S (2018) Environmental risk assessment of selected organic chemicals based on TOC test and QSAR estimation models. J Environ Sci 64:23–31

Dinis MDL, Fiúza A (2009) Methodology for exposure and risk assessment in complex environmental pollution situations. In: Exposure and risk assessment of chemical pollution—contemporary methodology. Springer, Netherlands, pp 111–132

Dong Z, Liu Y, Duan L, Bekele D, Naidu R (2015) Uncertainties in human health risk assessment of environmental contaminants: a review and perspective. Environ Int 85:120–132

Dubois D (2010) Representation, propagation, and decision issues in risk analysis under incomplete probabilistic information. Risk Anal: Int J 30(3):361–368

Eduljee GH (2000) Trends in risk assessment and risk management. Sci Total Environ 249(1–3):13–23

Elliot M, Thomas I (2009) Environmental impact assessment in australia: theory and practice. The Federation Press

EPA (2016). Human health risk assessment, 3 Feb 2021

Galatchi LD (2006) Environmental risk assessment. In: Chemicals as intentional and accidental global environmental threats, pp 1–6

Galimberti F, Moretto A, Papa E (2020) Application of chemometric methods and QSAR models to support pesticide risk assessment starting from ecotoxicological datasets. Water Res 174:115583

Gerba CP (2015) Risk assessment. In: Environmental microbiology, 3rd edn. Elsevier Inc., pp 565–579

Moy GG (2014) Risk analysis: risk analysis of hazards in food: an overview. In: Encyclopedia of food safety, vol 1. Elsevier

Nachman KE, Fox MA, Sheehan MC, Burke TA, Rodricks JV, Woodruff TJ (2011) Leveraging epidemiology to improve risk assessment. Open Epidemiol J 4:3–29

Pellizzato F (2014) Environmental risk assessment, pesticides and biocides. In: Encyclopedia of toxicology, 3rd edn. Elsevier, pp 402–405

Renn O (2008) Risk governance: coping with uncertainty in a complex world, vol 1. Earthscan, London

Swaen GM (2006) A framework for using epidemiological data for risk assessment. Hum Exp Toxicol 25(3):147–155

Swartjes FA (2015) Human health risk assessment related to contaminated land: state of the art. Environ Geochem Health 37(4):651–673

U.S. EPA (U.S. Environmental Protection Agency) (2000) The risk characterization handbook. United States Environmental Protection Agency

U.S. EPA (U.S. Environmental Protection Agency) (2022a) Human health risk assessment. https://www.epa.gov/risk/human-health-risk-assessment

U.S. EPA (U.S. Environmental Protection Agency) (2022b) Basic information about the integrated risk information system. https://www.epa.gov/iris/basic-information-about-integrated-risk-information-system

Zeleňáková M, Zvijáková L (2017) Risk analysis within environmental impact assessment of proposed construction activity. Environ Impact Assess Rev 62:76–89

Zhang S, Han Y, Peng J, Chen Y, Zhan L, Li J (2023) Human health risk assessment for contaminated sites: a retrospective review. Environ Int 171:107700

Chapter 11
Introduction to Strategic Environmental Assessment (SEA)

11.1 Introduction

A strategic environmental assessment, or SEA, is a thorough and systematic procedure for assessing the potential effects of proposed policies, programs, or plans on the environment. It serves as a means to address cumulative effects and takes into account economic and social aspects to facilitate well-informed decision-making (Sadler et al. 2012). Defining Strategic Environmental Assessment (SEA) presents a multifaceted challenge, with numerous proposed definitions. Initially, SEA was perceived as an extension of project-level Environmental Impact Assessment (EIA) to higher tiers of decision-making. An early definition by outlined SEA as a structured and all-encompassing process for evaluating how plans, policies, and programs will affect the environment; this includes compiling a thorough report and using the results to inform decisions. It is a procedure that seeks to include the environmental component in the planning and developing of programs, policies, and plans (PPPs). By creating environmentally sound strategies, this integration ensures that any operations and projects that are subsequently accepted, along with any other type of general environmental intervention, won't negatively impact the environment (Souloutzoglou and Tasopoulou 2020). Subsequent definitions, such as the one by Sadler and Verheem (1996), underscored SEA as a systematic approach for evaluating the environmental implications of proposed initiatives at the earliest suitable stage of decision-making, alongside economic and social considerations. To encapsulate the continuous, forward-thinking, and holistic nature of SEA, Partidário (1999) emphasized SEA as a process rather than a mere report production. This viewpoint stressed the adaptive and incremental aspects of SEA, its wide-ranging scope encompassing sustainability concerns, and its place more emphasis on ideas and visions than on particular acts and results.

The relationship between SEA and EIA has been a subject of ongoing discourse since the inception of SEA. The primary objectives of SEA encompass the evaluation of a policy, plan, or program in the following key areas:

© The Author(s), under exclusive license to Springer Nature Switzerland AG 2024
R. Bhateria et al., *Environmental Impact Assessment*,
https://doi.org/10.1007/978-3-031-66797-8_11

1. Responsiveness to environmental and climate change challenges: SEA assesses how effectively the proposed plan, policy, or program addresses environmental and climate-related issues.
2. Identification of potential harmful consequences on the environment and climate resilience: SEA aims to identify and evaluate any negative impacts that the proposed initiative may have on the environment and its ability to withstand climate change.
3. The examination of prospects for environmental improvement and the support of climate-resilient and low-carbon development: SEA aims to uncover possibilities for enhancing the environment and fostering sustainable, resilient, and low-carbon development within the proposed policy, plan, or program (Dalal-Clayton and Sadler 1999).

For optimal outcomes, integrating SEA into the early stages of policy, plan, or program development is essential, and it requires strong commitment and ownership from the government. Furthermore, active public participation is crucial for the success of SEA. The goals and objectives of SEA include evaluating environmental effects, identifying the most environmentally beneficial options, providing early warning of cumulative effects, streamlining project-level EIA by addressing strategic concerns, determining possible effects and information needs, and cutting down on the time and effort needed for individual assessment (Partidário 2000).

While defining SEA can be complex, it primarily entails a comprehensive environmental evaluation of policies, plans, and programs. Early definitions of SEA focused on its formalized, systematic, and comprehensive nature, involving the evaluation of environmental impacts, preparing written reports, and utilizing the findings for decision-making purposes. Over time, the concept of SEA evolved to emphasize its proactive, integrated, adaptive, and ongoing nature. It also expanded to encompass sustainability issues and shifted its focus from concrete actions and outcomes to visions and initiatives. The sustainability mandate of SEA has also been supported by academic literature. "The main rationale for applying SEA is to help create a better environment through informed and sustainable decision-making," states Fischer (2003). Sustainability is fundamental to SEA, according to Arce and Gullón (2000), and according to Linacre et al. (2006) and Liou and Yu (2004), SEA enhances governing by educating decision-makers on the sustainability of strategic actions. The limitations of project-level Environmental Impact Assessment (EIA) have necessitated the development of SEA as a drive to address the complexities of environmental decision making on a global scale. Several factors contributed to the need for SEA:

1. Timing of decisions: Project-level EIA often occurs too early in the decision-making process to consider the critical environmental impacts of policy and planning decisions made without a systematic impact assessment process (Von Seht 1999).
2. Nature of decisions: The Characteristics of Decisions: Policy and planning decisions are frequently gradual and indistinct, rendering them challenging to address using the conventional project-level Environmental Impact Assessment (EIA)

process. Consequently, a more strategic tool for assessing their impact was required.

3. Information availability: Frequently, there is limited information available at the policy and planning level, resulting in uncertainties that impede the detailed data demands of project-level Environmental Impact Assessments (EIA).

   3.1. Sustainable Development and Environment Protection:

   - Examining how suggested strategic initiatives will affect the environment.
   - Identifying the most practical environmentally friendly approach.
   - Offering early detection of cumulative impacts and substantial alterations.

   3.2 Enhancing and Streamlining Project Environmental Impact Assessment (EIA):

   - Preemptively describing the range of possible effects and information requirements.
   - Addressing strategic problems and considerations concerning proposal justifications.
   - Reducing the time and resources required for individual evaluations.

   3.3 Incorporating Environmental Considerations into Sector-Specific Decision-Making:

   - Advocating for eco-friendly and sustainable proposals.
   - Catalyzing a shift in the decision-making process.

11.2 The Evolution of Strategic Environmental Assessment (SEA)

The roots of Strategic Environmental Assessment (SEA) can be tracked down to the enactment of the National Environmental Policy Act (NEPA), which laid the initial groundwork for what would eventually evolve into SEA. NEPA introduced the concept of Environmental Impact Statements (EIS) and mandated federal agencies to prepare them for "major federal actions that significantly affect the quality of the human environment, including legislation" (National Environmental Policy Act (NEPA) of 1969, Section 102(2)(c)).

The history of environmental assessment in Western nations took shape in 1969 with the passage of the NEPA by the United States Congress (Finnveden et al. 2003). In 1985, the European Council introduced Directive 85/337/EEC, which established the framework for environmental assessment in Europe. The primary aim of this framework was to ensure that decision-makers had access to comprehensive and precise information regarding potential environmental and human health risks when considering permits for activities with potential environmental impacts. Within this framework, individuals or organizations intending to undertake such activities were required to develop an Environmental Impact Report (EIR). The EIR was expected to outline the measures taken or planned to minimize environmental impacts and demonstrate that the selected technologies and methods were the safest and most appropriate among the available choices. These reports were subject to discussion with relevant environmental protection authorities and those who might be affected by associated risks. Subsequently, the complete set of information was presented to decision-makers or governing bodies responsible for evaluating the advantages of the proposed activities in contrast to the potential environmental harm. Based on this assessment, a decision was reached, either granting or denying the permit for the proposed activity. The decision and its rationale were then made accessible to the public. The implementation of SEA has evolved over time, with NEPA and the European Directive serving as important milestones in its development. These frameworks aimed to ensure that environmental considerations were integrated into governing processes, providing a comprehensive understanding of potential risks and impacts. By adopting SEA, countries sought to enhance environmental protection and achieve sustainable development by systematically assessing the environmental consequences of plans, programs, and policies.

11.3 The Role and Benefits of Strategic Environmental Assessment (SEA)

It is crucial to emphasize that Strategic Environmental Assessment (SEA) should not be regarded as a mere replacement for the limitations of project-level EIA or as a solution to the challenges associated with implementing and understanding project-level EIAs (Von Seht 1999). SEA could not be seen solely as a tool for comparing and evaluating alternative projects, nor should project-level EIAs be reduced to the sole task of devising mitigation measures, when a project-level EIA is unable to adequately compare and evaluate the effects of real project alternatives, to find practical mitigation strategies, encourage public involvement, and guarantee appropriate monitoring, it is not able to properly inform decision-making, it does not necessarily imply that SEA is the remedy for these issues (Partidário 2003). Unfortunately, this misconception often persists. Many environmental assessment approaches currently categorized as SEAs may raise questions about their truly strategic nature. Determining whether an environmental approach falls under SEA or project-level EIA, or if it simply represents improved environmental planning practices, can be challenging. When more effective management or more proven methods of environmental assessment, such as project-level EIA, can adequately address the circumstances, introducing new approaches like SEA may not be necessary unless clear additional benefits are evident.

Nevertheless, SEA can genuinely enhance the decision-making process when it positively affects the standard of ensuing development initiatives. This is feasible due to SEA:

- Facilitates the integration of sustainability principles into decision-making
- Organizes environmentally informed actions in a tiered manner
- Provides a better context for evaluating cumulative effects
- Offers a screening framework for lower-level Environmental Assessments, particularly project-level EIAs
- Foresees potential impacts at the project level, thereby augmenting and fortifying project-level EIA.

Most practitioners see SEA as a tool for planning that can be employed flexibly at different points in the policy-making cycle, as opposed to a decision-making process (like EIA). From this wide angle, SEA includes evaluations of more specific plans and programs with physical and spatial links, as well as larger policy initiatives (e.g. local and metropolitan plans, and regional development programs). This coverage's breadth makes one issue very obvious. The approaches that should be used at the two extremes of the making choices spectrum are very different (Dalal-Clayton and Sadler 1999). SEA and project-level EIA are distinct processes, and their differences have been extensively discussed as a key rationale for the development of SEA (Table 11.1.).

The EIA is improved by the SEA. The distinction between the two is that a SEA is completed earlier than an EIA. The distinction between the two is that a SEA is completed earlier than an EIA. A SEA is conducted at the planning stage because significant environmental decisions are frequently made in the framework of preparatory strategies and initiatives, but an EIA is not conducted until an ecologically important project approaches the reviewing process. The Fig. 11.1 illustrates how policies, plans, and programs are often idealized as a hierarchical method for decision-making. But in practice, creating policies doesn't always follow a rational flow of distinct, technical procedures. It is further an intricate, repetitive process where the alternatives are steadily reduced and the majority are closed off by the project stage (Dalal-Clayton and Sadler 1999).

Policies are seen to be large declarations of schemes that serve to both foci and reflect a government's political agenda and start a decision-making cycle. Plans and programs, which include choosing options to attain policy objectives and outlining how, where, and when specific activities will be executed, give them substance and effect (Sadler and Verheem 1996).

Table 11.1 Distinctions between EIA and SEA

EIA	SEA
It responds to a development proposal	It takes a proactive stance and provides information to shape development proposals
It is employed to evaluate the impact of a proposed development on both the environment and socio-economic conditions	It is utilized to evaluate how existing environmental and socio-economic conditions affect development opportunities and limitations
It pertains to a particular project	It pertains to specific areas, regions, or sectors of development
It allows for the identification of impacts that are specific to the project in question	It facilitates the establishment of a framework for measuring both positive and negative impacts
It follows a clearly defined start and finish and is geared towards providing information for a certain decision at a juncture	It is the process of developing a framework for sustainability to direct continuous decision-making over a long time
It places its focus on mitigating negative impacts and amplifying positive ones	It concentrates on preserving a specific level of environmental quality and socio-economic conditions, often achieved by identifying sustainable objectives and defining acceptable change limits
It requires a great degree of detail and a limited viewpoint	It adopts a broad viewpoint and incorporates limited detail to offer a vision and overarching framework

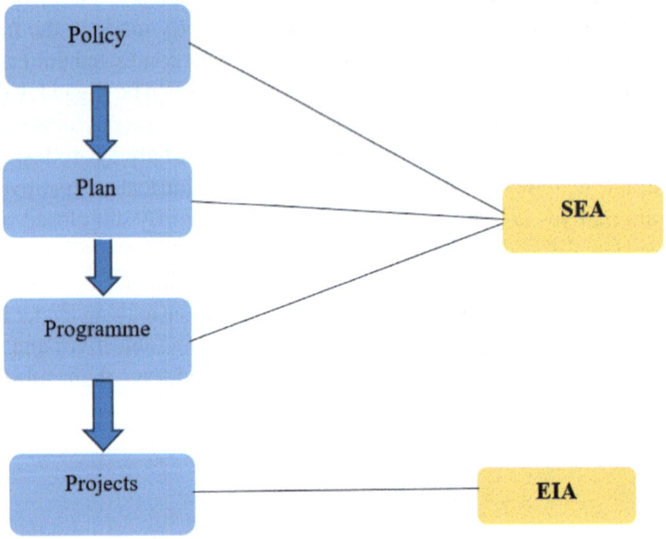

Fig. 11.1 Relation between SEA and EIA

11.4 Optimizing the Usefulness of Connections Between SEA and EIA

11.4.1. Connect program SEA and project EIA (in which tiering is applicable) explicitly. Program/project SEA may be linked if the project is large enough to meet both SEA and EIA requirements.
11.4.2. Parallel procedures may be required when both SEA and EIA are required because of local legal requirements.
11.4.3. Joint procedures may be used when SEA and EIA requirements can be satisfied by a single procedure when timings and scale align.

11.5 Steps in SEA

A well-defined set of methods and a precise process are necessary for SEA. The necessary steps encompass describing the state of the environment as it is at the moment, looking at potential alternatives, identifying, characterizing, and assessing any major effects on the environment, and incorporating the findings of the environmental evaluation and public hearings into the program or plan that is being approved. And lastly, the observation of any lingering effects following their execution (Iyer 2018).

Screening

Screening involves assessing whether the proposed action will negatively impact sustainable development and the environment, either directly or indirectly. The first stage of the SEA method is to decide whether it is necessary. There are several aspects to consider before making this decision. Before planning program initiators, or hired development bodies, must first confer with environmental and healthcare authorities. These authorities will then determine whether the SEA procedure (screening) is necessary for the particular plan or program. Generally, there are two methods to proceed with the screening process:

i. Examining each case individually and
ii. Defining the kinds of plans and programs that are created and approved under national law
iii. Combining the above two methods

Scoping

The second step in the SEA process is scoping, which entails determining the assessment's scope. While specific requirements for the scoping procedure or content are not mandated, it is recommended as a necessary step. Scoping aids in focusing the SEA on essential aspects, avoiding the production of irrelevant data and establishing the level and extent of detail of the assessment. Scoping involves identifying baseline data and information needs, determining sources and gaps in the current information, and describing environmental concerns and protection goals. Additionally, it is important to identify various development alternatives that can meet environmental, economic, and social objectives (Polido and Ramos 2015).

Impacts Assessment

Assessing the impacts of strategic actions is a critical component of SEA. Although the assessment of strategic proposals involves a higher level of uncertainty compared to project-level assessments, evaluating the likely effects of the proposed actions remains necessary. The primary goal of SEA is to select the preferred strategic actions that achieve objectives with minimal cost or maximum benefit to the environment. The assessment should consider explicit environmental criteria and relevant aspects such as probability, magnitude, reversibility, duration, and more. The indirect and cumulative consequences of several activities should receive further consideration. Prediction and evaluation can employ various techniques, including expert opinion, overlay maps, modeling, carrying capacity analysis, public consultation, and scenario analysis (Acharibasam and Noble 2014).

Mitigation

The goal of mitigation measures is to improve the suggested strategy's environmental performance. Although not a separate step in the SEA procedure, mitigation strategies ought to be taken into account at all times. SEA acts as a facilitator, suggesting mitigation measures to improve or remediate potential positive and negative environmental impacts identified during the assessment (Tinker et al. 2005).

Report Preparation

The environmental report presents the outcomes of the SEA process, including the assessment of the presented action and its expected impacts on the environment. The report serves as a basis for public participation, consultation, and decision-making. It must include a description of the proposed action and its alternatives, a description of the baseline environment, significant environmental impacts, timescales and likelihood of predicted impacts, potential mitigation measures, and comments on assessment issues. The report must meet the following fundamental criteria: it must be a concise, easily understood, and accessible document, with the majority of its content devoted to a succinct explanation of the anticipated effects of implementing the suggested plan or program, conclusions, and recommendations. The incorporation of environmental factors into decision-making processes is facilitated by clear documentation and report accessibility (Von Seht 1999). At present, reports frequently have what is known as a "website structure," even though the topics of interest are highly detailed. Another name for it is the pyramid structure.

Review

Review stage evaluates the quality of the information presented in the SEA process, ensuring its effective execution. During the review, the acceptability of collected and presented information about the environment in the report is examined. Uncertainties, contradictions, and biases should be identified and addressed. The recommendations derived from The SEA process serve as the foundation for incorporating environmental factors into strategic action decision-making.

Consultation and Public Participation

The implementation of a strategy is ensured by public participation in its implementation. Public participation and consultation and are crucial for establishing stakeholder ownership of the development, improving decision-makers understanding of issues, and increasing transparency in the strategy preparation process. While not separate stages, involving relevant authorities and the public at different points throughout the SEA process is ideal. Consultation and public involvement can occur during plan development, screening, scoping, impact assessment, report preparation, and the review stage. The SEA Directive emphasizes the importance of providing early and effective opportunities for authorities and comments on the proposed program or plan and the environmental report that goes with it from the general public (Rega and Baldizzone 2015).

Monitoring

Monitoring is essential to ensure that the strategic actions implemented through SEA stay on track to achieve their intended purpose. It involves periodic follow-up and a comparison of assessed performance and results against the plans. Monitoring should be flexible, considering that environmental impacts may not occur immediately and can arise from future consequential actions. It includes using selected indicators to assess performance and procedural quality. Policy review is also part of the monitoring phase, enabling the evaluation and adjustment of policy actions based on their outcomes. Monitoring data may be utilized to determine the effectiveness of mitigation strategies and SEA recommendations for protecting the environment, as well as whether they have been implemented. At the planning as well as project implementation levels, monitoring can assist in identifying unexpected consequences (and encourage prompt preventive action), addressing assessment uncertainties, and filling in data gaps found during the SEA. (Arts and Morrison-Saunders 2004; Gachechiladze-Bozhesku and Fischer 2012; Azcárate et al. 2013). The SEA process goes through five important stages starting with identifying and setting your goals, Strengths, Weaknesses, Opportunities, and Threats (SWOT) Analysis, forming strategy, executing strategy, and evaluating strategy as described in the flow diagram (Fig. 11.2).

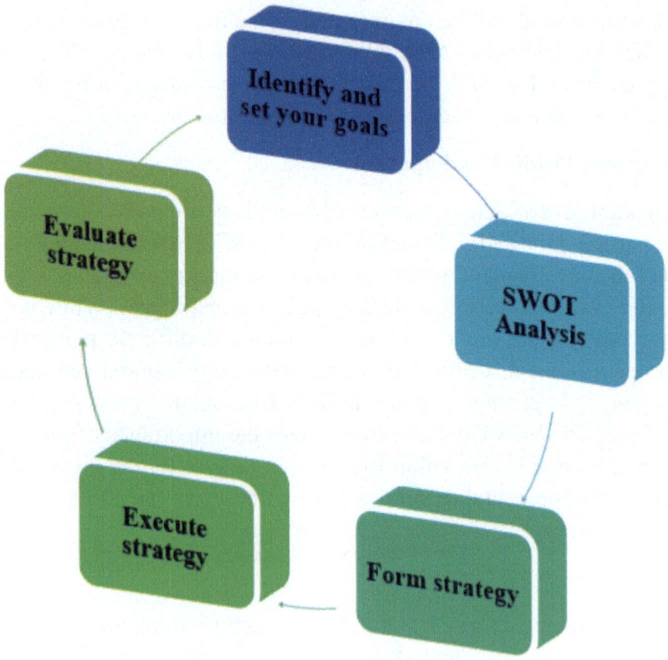

Fig. 11.2 Five general stages of strategic environmental assessment

11.6 Models and Methods of Strategic Environmental Assessment (SEA)

SEA methodologies are rooted in similar techniques and tools of Environmental Impact Assessment (EIA). However, SEA represents a new approach that focuses on the concept, scope, breadth of policy assessment, planning, and timing. The significance and strengths of each procedural approach in national and regional decision-making influence the adoption of SEA models (Nautiyal and Goel 2021). As a result, Methodological and procedural components of SEA techniques are frequently drawn from project-level EIA practices (bottom-up approaches) or policy assessment practices (top-down approaches) (Souloutzoglou and Tasopoulou 2020). The establishment of SEA generative models and techniques is based on two primary models.

11.6.1 Policy-Based Approach

This model follows a planning or policy rationale and incorporates environment assessment principles into the development of plans and policies. It involves identifying development needs and options that are then systematically assessed about sustainable development. Examples of this approach include the Resource Management Policy in New Zealand, the E-test in the Netherlands, the Policy Impact Assessment in Canada, and the SEA of government bills in Denmark.

11.6.2 Project-Based Approach

With this paradigm, plans and programs are included in the project-level EIA process. It makes use of the current EIA regulatory guidelines and processes in addition to the standard procedures of incorporating environmental data, techniques, alternative comparisons, and mitigation strategies. Programmatic environmental impact assessments in the US and SEIA in the Netherlands are two instances of this methodology. This is referred to as the standard (EIA-based) model by Sadler and Verheem (1996). Figure 11.3 shows two primary approaches/models for the development and reinforcement of SEA.

The general ailments (Fischer 2007; Therivel and Wood 2005) state that SEA procedures and methods ought to:

(1) Be appropriate for their intended use, which means they should be able to handle important problems and participate in the decision-making process.
(2) Be on the schedule for making decisions.

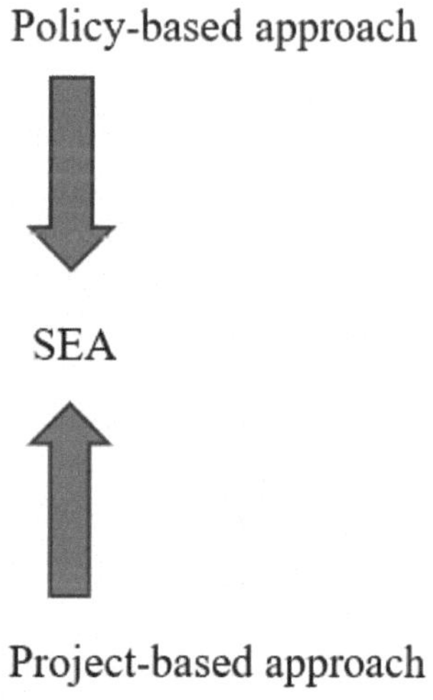

Fig. 11.3 Two primary approaches for the development and reinforcement of SEA

(3) Permit the blending of different substantive elements, such as different industries, offices, and processes.
(4) Be sure to account for uncertainty.
(5) Be strong, clear, and pertinent to the goal and the procedures.
(6) Be comprehensible to every SEA participant.
(7) Be economical.
(8) Offer alternatives, as well as methods to mitigate any negative consequences.
(9) Permit the possibilities to be compared.

These requirements should be viewed more as broad guidelines than as standards to be followed when choosing the best SEA approaches and strategies for a given situation. Therivel 2004 states that the following factors define which of these are most appropriate for a given situation: the scale (international, national, regional, local), the "strategicness" (policy, plan, program) of the strategic actions, the decision-making process, the audience for whom the findings are intended, the context in which SEA is carried out, the accessibility of time, resources, employees, and tools, as well as the kinds of data required by the methods and techniques. Last but not least, the customs and perspectives of any state, investigator, or practitioner have a common characteristic (Souloutzoglou and Tasopoulou 2020).

11.7 Methods and Techniques Used in SEA

SEA employs a diverse array of methods and techniques that draw from Environmental Impact Assessment (EIA) and policy analysis/planning (Liou et al. 2006; Brown and Therivel 2000). Some of the commonly used methods include:

- Checklists
- Input–output models
- Geographical Information Systems (GIS)
- Matrices
- Networks
- Scenarios and simulations
- Forecasting
- Analysis of land suitability
- Systems modeling
- Multi-criteria evaluation
- Objectives accomplishment matrices
- Creating plans for balance sheets
- Cost–benefit evaluation
- Techniques for reducing expenses
- Sensitivity evaluation

These methods collectively enable a comprehensive evaluation of the environmental and strategic aspects of proposed actions. The matrices and expert judgment were the most often utilized approaches and strategies. The first was primarily applied during the process of screening and scoping phases, while the second was applied during the identification, assessment, and evaluation of the direct, indirect, and cumulative effects as well as during the formulation of mitigation strategies. During the process of determining the scope and depth of implementation, the participatory approaches were largely employed among the remaining techniques and procedures. (Souloutzoglou and Tasopoulou 2020).

11.8 SEA Procedural Methodologies

When approaching SEA, several key questions should be addressed from the outset. These questions provide a basic framework for the SEA process:

1. Objectives and Purpose: What are our objectives, where do we want to go, and why?
2. Exploring Alternatives: What are the options available to achieve the same goal?
3. Sustainability Assessment: What may be the future consequences within a sustainability framework?
4. Integrated Decision-Making: How can we integrate environmental, social, and economic considerations into decision-making processes?

From an operational standpoint, these questions have crucial aspects that need to be addressed at various stages of a strategic-level impact assessment. They might be seen as a sequence of planned actions carried out using an integrated approach during the decision-making process or a sequential SEA process. The SEA procedural approach will be shaped and the order of actions to be carried out will be affected by various factors such as the timeline, resources available, legislation, strategy, or program building stage, and the political context of how decisions are made. (González 2022).

11.9 Methods

11.9.1 Future Studies

SEA often includes future considerations to improve decision-making. It commonly uses future scenarios, scenario analysis, forecasting, and backcasting techniques. Various approaches exist for studying the future, such as forecasts, conditional forecasts, scenarios, external scenarios, and policy scenarios. Forecasts extrapolate trends and mechanisms from historical data to project probable futures, while conditional forecasts combine extrapolations with conditional assumptions to generate multiple scenarios. In situations where there is a great deal of qualitative ambiguity regarding the future, scenarios can be quite helpful. External situations account for factors beyond the user's control but are still relevant to the study (Finnveden et al. 2003). Policy scenarios come into play when the user can actively influence the future rather than merely reacting to it. Backcasting involves defining a desired future goal and then envisioning pathways to achieve it. Modeling can support different future study approaches, relying on assumptions based on historical trends or incorporating new mechanisms. Quantitative information can be generated using models, but complex models may sacrifice transparency. Dynamic modelling, such as the MARKAL and MARTES models, is commonly employed in energy systems studies.

11.9.2 Life Cycle Assessment (LCA)

LCA is a procedure for evaluating the environmental effects and resource consumption across the animateness of a product, encompassing stages from primal matter procurement and manufacturing to use and dumping. For LCA, a standard by ISO has been set up, providing language, structure, and methodology choices. The assessment comprises four phases including setting objectives and parameters, making an inventory, carrying out a life cycle impact assessment (LCIA), and deriving conclusions. A thorough environmental assessment is given by LCA, yet not all environmental aspects are uniformly fully covered. Land use, toxicological effects, radiation, accidents, and disamenities are often challenging to assess within LCA. LCA traditionally provides site- and time-independent results, but there is a trend toward making it more site-dependent or even site-specific. Weighting methods can be employed to aggregate indicator results across impact categories, including economic valuation and multi-attribute approaches. LCA influences the pressure and impact aspects within the DPSIR framework by offering data on emissions and their potential consequences. Nevertheless, it doesn't portray the comprehensive state as it exclusively focuses on emissions directly associated with the assessed functional unit. (Finnveden et al. 2003).

11.10 Advantages of Strategic Environmental Assessment (SEA)

1. Enhances public participation and consultations
   - Encourages public engagement and involvement in the governing process.
2. Identifies objectives and potential harm to ecosystems and human being health
   - Assesses and determines the objectives and likely consequences related to environmental and public health aspects (Gazzola et al. 2019).

3. Considers scenarios and conducts screening
   - Evaluate different scenarios and conduct screening to identify potential alternatives and options.
4. Sets targets and identifies alternatives
   - Establishes targets and explores alternative solutions to align with environmental and sustainable development goals.
5. Supports environmentally sound and sustainable development
   - Promotes sustainable development by giving equal importance to environmental, social, and economic aspects.
6. Facilitates coordination and reconciles different objectives
   - Helps in coordinating efforts among various administrative levels and sectors, balancing diverse objectives.
7. Addresses the root causes of environmental problems
   - Takes a proactive approach to address the underlying causes of environmental issues, rather than just treating the symptoms.
8. Strengthens policy, plan, and program-making processes
   - Enhances the effectiveness of policy, plan, and program formulation by considering environmental factors from the outset.
9. Saves time and money
   - Identifies unsustainable development options early on, saving resources by avoiding costly investments in problematic projects.
10. Improves public trust and governance
    - Enhances transparency in governing, builds public trust, and incorporates stakeholders' opinions in the planning process.
11. Preserves a healthy environment and reduces remediation costs
    - Promotes environmental preservation, reducing the need for expensive remediation measures in the future.
12. Enhances credibility and stakeholder support
    - Improves the credibility of policies, plans, and programs, gaining support from key stakeholders for their implementation (Desmond 2007).

Overall, SEA provides a range of immediate and long-term benefits, including improved governance, sustainable development, cost savings, and public trust in decision-making processes (Partidário 2000) Effective implementation of SEA can be achieved through integrating society, environment, and economy (Fig. 11.4).

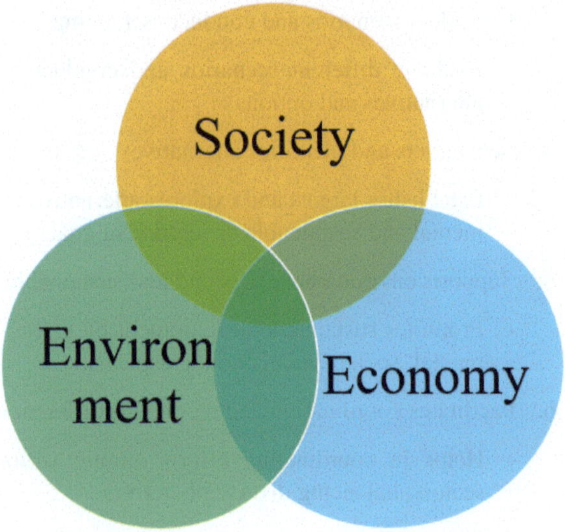

Fig. 11.4 Integration of environment, economy, and society for effective Strategic Environment Assessment

Strategic environmental assessment (SEA) is a systematic process for evaluating the environmental consequences of proposed policies, plans, or programs, integrating environmental considerations into decision-making. While traditionally associated with land-use planning and infrastructure development, SEA can be applied across various fields where there are significant interactions between human activities and the environment (Gazzola et al. 2019). Some of these fields include urban planning, agriculture planning, industrial development, natural resource management, energy sector, and many more (Fig. 11.5).

11.11 The Principles of Applying the SEA Process Effectively

To effectively formulate, revise, and implement policies, plans, and programs, as well as to influence decision-making, a certain level of organizational maturity is required. Appropriate skills are also required within government programs and institutions, in the private sector (such as industry, and environmental consulting companies), among academics, and in non-governmental organizations. Finally, these sectors must have sufficient ability (for both human beings and financial assets) (Sadler and Baxter 1997). Six principles are described for effective implementation of SEA.

(1) Preliminary SEA will be applied to all plans and programs whose implementation could have a major impact on the nation's sustainable development
(2) The goal of SEA is to fully inform stakeholders about the risks associated with implementing a plan or program for sustainable development and whether or not there are less risky options

Fig. 11.5 Applicability of SEA in various sectors

(3) To enable the inclusion of its findings in the plans and programs themselves, SEA must be carried out as early in the development process as feasible
(4) The appropriate and trustworthy method for quality assurance must be in place
(5) The appropriate system must be in place to educate decision-makers about the stances and viewpoints of all parties involved
(6) The ultimate judgment, assessment, and plan/program must be made public. Stakeholders must be notified that some of their viewpoints were taken into account and why, with explicit justification (Partidário 2003).

Summary

SEA serves as a comprehensive and systematic tool for integrating environmental considerations into decision-making processes, particularly within the realm of strategic planning and policy formulation. The evolution of SEA has showcased its adaptability and responsiveness to the dynamic challenges posed by rapid urbanization, industrialization, and globalization. From its origins as a project-level assessment tool, SEA has evolved into a strategic instrument that addresses the interconnectedness of environmental, social, and economic factors on a broader scale.

One of the notable strengths of SEA is its ability to foster stakeholder engagement and collaboration. By involving diverse stakeholders, including local communities,

NGOs, and industry representatives, SEA promotes transparency, inclusivity, and the incorporation of multiple perspectives into decision-making processes. This participatory approach not only enhances the quality of environmental assessments but also contributes to the legitimacy and acceptance of strategic plans and policies. Despite its achievements, challenges persist in the effective implementation of SEA, ranging from capacity constraints to the need for standardized methodologies. Continued research and development in these areas will be essential to further enhance the robustness and consistency of SEA processes worldwide. SEA's continued refinement, coupled with a commitment to collaboration and innovation, will undoubtedly contribute to a more resilient and sustainable future.

Questions/Exercise

1. How does SEA address the limitations of project-level Environmental Impact Assessment (EIA)? Compare both.
2. What role does public participation play in the success of SEA?
3. What are the key steps involved in the SEA procedure? How does monitoring in SEA, ensure the ongoing effectiveness of strategic actions and the protection of the environment?
4. What are the general ailments that SEA procedures and methods should address, and how do these requirements, contribute to effective decision-making in SEA?
5. What is the significance of having an appropriate method for quality assurance in the context of SEA implementation?

References

Acharibasam JB, Noble BF (2014) Assessing the impact of strategic environmental assessment. Impact Assess Proj Apprais 32(3):177–187. https://doi.org/10.1080/14615517.2014.927557

Arce R, Gullón N (2000) The application of strategic environmental assessment to sustainability assessment of infrastructure development. Environ Impact Assess Rev 20(3):393–402. https://doi.org/10.1016/S0195-9255(00)00050-0

Arts J, Morrison-Saunders A (eds) (2004) Assessing impact: handbook of EIA and SEA follow-up. Routledge, London

Azcárate J, Balfors B, Bring A, Destouni G (2013) Strategic environmental assessment and monitoring: Arctic key gaps and bridging pathways. Environ Res Lett 8(4):044033. https://doi.org/10.1088/1748-9326/8/4/044033

Brown AL, Therivel R (2000) Principles to guide the development of strategic environmental assessment methodology. Impact Assess Proj Apprais 18(3):183–189. https://doi.org/10.3152/147154600781767385

Dalal-Clayton DB, Sadler B (1999) *Strategic environmental assessment: a rapidly evolving approach*. International institute for environment and development, London. https://www.jstor.org/stable/resrep18001

Desmond M (2007) Strategic environmental assessment (sea): a tool for environmental decision-making. Ir Geogr 40(1):63–78. https://doi.org/10.1080/00750770709555887

Finnveden G, Nilsson M, Johansson J, Persson Å, Moberg Å, Carlsson T (2003) Strategic environmental assessment methodologies—applications within the energy sector. Environ Impact Assess Rev 23(1):91–123. https://doi.org/10.1016/S0195-9255(02)00089-6

Fischer TB (2003) Strategic environmental assessment in post-modern times. Environ Impact Assess Rev 23(2):155–170. https://doi.org/10.1016/S0195-9255(02)00094-X
Fischer TB (2007) Theory and practice of SEA: towards a more systematic approach. Earthscan, London
Gachechiladze-Bozhesku M, Fischer TB (2012) Benefits of and barriers to SEA follow-up—theory and practice. Environ Impact Assess Rev 34:22–30. https://doi.org/10.1016/j.eiar.2011.11.006
Gazzola P, Belčáková I, Pauditšová E (2019) Landscape impact assessment in planning processes. De Gruyter, p 200. https://doi.org/10.1515/9783110601558
González A (2022) Strategic environmental assessment monitoring: the enduring forgotten sibling. Impact Assess Proj Apprais 40(2):168–176. https://doi.org/10.1080/14615517.2022.2031552
Iyer VG (2018) Strategic Environmental Assessment (SEA) process towards sustainable construction management development for the electrical, automation, and mechanical engineering construction industries' excellence achievements. In: 2018 3rd international conference on electrical, automation and mechanical engineering (EAME 2018), June 2018. Atlantis Press, pp 87–96. https://doi.org/10.2991/eame-18.2018.18
Linacre NA, Gaskell J, Rosegrant MW, Falck-Zepeda J, Quemada H, Halsey M, Birner R (2006) Strategic environmental assessments for genetically modified organisms. Impact Assess Proj Apprais 24(1):35–43. https://doi.org/10.3152/147154606781765309
Liou ML, Yu YH (2004) Development and implementation of strategic environmental assessment in Taiwan. Environ Impact Assess Rev 24(3):337–350. https://doi.org/10.1016/j.eiar.2003.10.018
Liou ML, Yeh SC, Yu YH (2006) Reconstruction and systemization of the methodologies for strategic environmental assessment in Taiwan. Environ Impact Assess Rev 26(2):170–184. https://doi.org/10.1016/j.eiar.2005.08.003
Nautiyal H, Goel V (2021) Sustainability assessment: metrics and methods. In: Methods in sustainability science. Elsevier, pp 27–46. https://doi.org/10.1016/B978-0-12-823987-2.00017-9
Partidário MR (1999) Strategic environmental assessment—principles and potential. Handb Environ Impact Assess 1:60–73
Partidário MR (2000) Elements of a SEA framework—improving the added-value of SEA. Environ Impact Assess Rev 20(6):647–663. https://doi.org/10.1016/S0195-9255(00)00069-X
Partidário MR (2003) Strategic Environmental Assessment (SEA): current practices, future demands, and capacity-building needs. International Association for Impact Assessment IAIA Training Courses.
Polido A, Ramos TB (2015) Towards effective scoping in strategic environmental assessment. Impact Assess Proj Apprais 33(3):171–183. https://doi.org/10.1080/14615517.2014.993155
Rega C, Baldizzone G (2015) Public participation in strategic environmental assessment: a practitioners' perspective. Environ Impact Assess Rev 50:105–115. https://doi.org/10.1016/j.eiar.2014.09.007
Sadler B, Baxter M (1997) Taking stock of SEA. Environ Assess 5(3):14–16 (special issue on SEA)
Sadler B, Dusik J, Fischer T, Partidario M, Verheem R, Aschemann R (eds) (2012) Handbook of strategic environmental assessment. Routledge
Sadler B, Verheem R (1996) Conclusions and recommendations of the international SEA effectiveness study. In: Proceedings of the 16th annual meeting of the International Association for Impact Assessment (IAIA), Estoril, Portugal, 17–23 June 1996
Souloutzoglou A, Tasopoulou A (2020) The methods and techniques of strategic environmental assessment. Comparative evaluation of Greek and international experience. Sustainability 12(8):3310. https://doi.org/10.3390/su12083310
Therivel, R (2004) Strategic environmental assessment in action. Earthscan, London, UK
Therivel R, Wood G (2005) Tools for SEA. In: Schmidt M, Joao E, Albrecht E (eds) Implementing strategic environmental assessment. environmental protection in the European Union, vol 2, pp 349–362. Springer Science & Business Media, Berlin/Heidelberg, Germany

Tinker L, Cobb D, Bond A, Cashmore M (2005) Impact mitigation in environmental impact assessment: paper promises or the basis of consent conditions? Impact Assess Project Appraisal 23(4):265–280. https://doi.org/10.3152/147154605781765463
Von Seht H (1999) Requirements of a comprehensive strategic environmental assessment system. Landsc Urban Plan 45(1):1–14. https://doi.org/10.1016/S0169-2046(99)00026-2

SPRINGER NATURE

GPSR Compliance

The European Union's (EU) General Product Safety Regulation (GPSR) is a set of rules that requires consumer products to be safe and our obligations to ensure this.

If you have any concerns about our products, you can contact us on ProductSafety@springernature.com

In case Publisher is established outside the EU, the EU authorized representative is:

Springer Nature Customer Service Center GmbH
Europaplatz 3
69115 Heidelberg, Germany

The manufacturer's authorised representative in the EU is Springer Nature Customer Service Centre GmbH, Europaplatz 3, 69115 Heidelberg, Germany. If you have any concerns regarding our products, please contact ProductSafety@springernature.com

Printed and bound by CPI Group (UK) Ltd, Croydon, CR0 4YY
30/09/2026
02225607-0018